Shaktipat

The Doorway To Enlightenment

Mark Griffin

Hard Light Center of Awakening

Shaktipat © 2012 by Mark Griffin,
Hard Light Center of Awakening
ISBN 978-0-9819375-0-2

All rights reserved. No part of this book may be reproduced in any form or by any means, electronic or mechanical, including photocopying, recording, or by any information storage and retrieval system, without the express written permission from the author.

For more information about the **Hard Light Center of Awakening** please visit *www.hardlight.org*. The web site provides a complete listing of Mark Griffin's other books and CDs, as well as links to audiobooks, podcasts and PDF books.

The **Hard Light Center of Awakening** is an organization founded and directed by Mark Griffin as a forum for the study of spirituality and meditation. Mark Griffin is a Meditation Master who is firmly established in the advanced Nirvikalpa Samadhi states — rare strands of consciousness that lead to remarkable perception and spiritual accomplishment.
First Edition: March 2012

Transcription: Ann Brockway and Evelyn Jacob

Editing, Layout: Mindy Rosenblatt and Evelyn Jacob / PodPublishing

TABLE OF CONTENTS

The Siddha Lineage	vii
Introduction	ix
Foreword	1
CHAPTER ONE – WHAT	17
The Four Bodies	19
The Four States of Consciousness	26
The Pervading Nature of Prana	30
SoHam and Kundalini Awakening	36
Meditation Instructions	46
CHAPTER TWO – WHY	51
Samsara, The Wheel of Cyclic Existence	53
The Guru Principle and Shaktipat	56
The Siddha Lineage	75
Divine Will	84
CHAPTER THREE – HOW	95
The Guru Margena - The Path of the Guru	99
The Rupa ~ The Blue Pearl	109
The Structure of Sahasrara	112
Where Does This Lead?	127
Perceiving the True World & Emptiness	135
Recognition and Samadhi	143
CHAPTER FOUR – WHEN	147
Karma, Impressions and Grace	153
Drawing Grace Into Our Lives	164
Bringing Karma Into Equilibrium	170
Unspooling Karma and The Fourth State	176
GLOSSARY	195
ADDITIONAL RESOURCES	215
Other Books by Mark Griffin	217

THE SIDDHA LINEAGE

Mark Griffin is a Westerner who was born in the 1950's in the Pacific Northwest. His early adult years were spent in the aggressive pursuit of higher knowledge and purpose.

While a young man, Mark's studies in art and music brought him to the San Francisco Bay area. There in 1976 he met his Guru, Baba Muktananda. After years of full-time immersion in the study of meditation, Mark encountered a milestone of extreme spiritual significance – entrance into the advanced state of consciousness known as Nirvikalpa Samadhi. After Muktananda died, Mark continued to study with the great teachers of the Kagyu tradition, Kalu Rinpoche and Chogyam Trungpa, who supported the maturing and stabilizing of his abilities.

In 1989, after attracting several interested students, Mark began to teach meditation. He and his students relocated to Los Angeles and started the Hard Light Center of Awakening, an association dedicated to the art and science of awareness of the Self.

When Mark Griffin met Baba Muktananda he immediately realized that Baba was his Guru, his true teacher. Baba's Guru was the great saint of India, Bhagawan Nityananda of Ganeshpuri. It is with the blessings of these remarkable Siddhas that Mark carries on his inspiring teachings.

INTRODUCTION

For a chemist, the word catalyst means a substance that rapidly brings about change, allowing transformations to occur that may not have been possible without the introduction of that catalytic agent.

For a spirtual seeker, the concept of shaktipat is understood in the same way. It is a moment of pure contact with the divine that quickens the journey towards awakening and raises the state of the seeker to an entirely new level of understanding and experience.

In this volume of the Essential Spiritual Training series, we focus on shaktipat as offered and explained by meditation master Mark Griffin.

We have selected four talks for this book that Mark gave between July 2007 and February 2009 because they answer four of the most salient questions a seeker can ask:

What is shaktipat and what is its role in my sadhana, my spiritual journey?

Why should I be interested in receiving shatkipat?

How is shaktipat given – what is the Guru really doing when he confers this transmission?

When does shaktipat become available to me and why am I interested in it now?

Of course, Mark discusses spiritual training more broadly than answering just these questions, so we've included the complete teaching from each associated session.

We offer our sincere thanks to Professor Patrick Mahaffey for his brilliant foreword, and to the Guru Tattva – the Guru principle – which exists in us all and is so strongly embodied in Mark Griffin and the great beings of the Siddha lineage.

May All Beings Awaken

Mindy Rosenblatt & Evelyn Jacob

PodPublishing

FOREWORD

We live in a time when many in the West seek to undertake the great journey of spiritual awakening. Success in this endeavor requires a reliable guide and a map of the territory to be traversed. This book on yoga spirituality by a contemporary Western meditation master is a boon for seekers aspiring to discover their true nature.

Mark Griffin is a spiritual teacher or Guru who transmits the teachings and practices of yoga masters known as siddhas, spiritual adepts who have fully realized their true nature and who have the capacity to transmit the light of awakening to sincere spiritual seekers. Siddhas embody the teachings they convey to others and empower seekers to realize the truth they teach in their own lives.

The process of awakening is an experiential matter, a path one travels rather than a matter of belief. Nevertheless, this process can be more readily understood and appreciated if it is situated in the context of yoga spirituality and the vocabulary that expresses yoga's esoteric principles and practices.

All great religions or wisdom traditions contain contemplative teachings and practices. Yoga is the essence of the religious and mystical traditions that originated in India. Since the 1960s, meditation masters from India have transmitted teachings and practices of yoga to receptive seekers in the West who yearn for inner peace, genuine happiness, and an experience of the divine.

Swami Muktananda was one of these great masters. He sought to create a "meditation revolution" by awakening the kundalini, the dormant spiritual energy of those seekers who came to see him during his three world tours in the 1970s and early 1980s. Mark Griffin was among the earliest of Muktananda's Western disciples to receive shaktipat initiation in 1976. After years of intense sadhana or spiritual practice, Griffin entered the profound state of nirvikalpa samadhi and awakened to his true nature. After Muktananda died, Griffin continued to study with Kalu Rinpoche and Chogyam Trungpa, great Tibetan teachers of the Kagyu tradition of Vajrayana Buddhism. He subsequently began to teach and to transmit shaktipat to sincere seekers, establishing the Hard Light Center of Awakening in 1989 as the vehicle for guiding his students through weekly study and practice sessions, day-long meditation intensives, and retreats.

Today more than 20 million Americans currently practice yoga postures to cultivate flexible bodies and physical well-being. However, yoga postures or asanas are only one aspect or limb of the philosophy of yoga systematically described in the Yoga Sutras. In this classical work, Patanjali describes the eight limbs of integrated practice. The first five limbs pertain to the outer aspect of yoga: ethical conduct (yama), spiritual discipline (niyama), posture (asana), breath control (pranayama), and withdrawal of the senses (pratyahara). These limbs enable a practitioner to shift one's attention from the outer world of sense objects to the inner world of meditation.

The last three aspects of practice describe how this occurs: the yogi concentrates his or her attention on a single object such as the breath or a mantra (dharana); concentration gives rise to one-pointed attention on the object (dhyana); finally, attention is fully absorbed into the object (samadhi). Samadhi is the stage of meditation in which the mind is completely concentrated and a superconscious mode of knowing comes into play.

The aim of yoga, according to Patanjali's text, is to still the thoughts and emotions in the heart-mind that afflict our lives and obscure our awareness of our true nature. Thoughts and emotions are described as fluctuations or modifications of the heart-mind animated by memories and energies stored in the unconscious. Yoga is the means for quieting these forces so that the pure awareness that underlies the mental modifications may shine forth and illumine the mind. Sustained practice confers freedom from the limitations of conditioned experience. Put another way, the purpose of practice is to discern a reality beneath the emotional and intellectual convolutions that distort our ordinary egoic awareness. Nothing less is the true aim of yoga.

Yoga is a multifaceted phenomenon. It refers both to traditions of mental and physical discipline and the goal achieved by those disciplines. Most frequently it is interpreted as the union of the individual Self (jiva-atman) with the supreme or universal Self (parama-atman). This meaning is derived from the Sanskrit root for yoga, yuj, meaning "to unite, join, connect." For Patanjali, compiler of the aforementioned Yoga Sutras, yoga entails the focusing of attention on whatever object is being contemplated to the exclusion of all others.

Georg Feuerstein provides the best comprehensive definition: "In its technical sense, yoga refers to that enormous body of spiritual values, attitudes, precepts, and techniques that have been developed in India over at least five millennia and that may be regarded as the very foundation of the ancient Indian civilization. Yoga is thus the generative name for the various Indian paths of ecstatic self-transcendence—the methodical transmutation of consciousness to the point of liberation from the spell of the ego-personality."[1] The plethora of paths reflects both the innovations and the particular concerns of the various texts relevant to the social context for engaging in yogic practice.

The earliest phase of yoga may be traced to the Vedas and Upanishads, the most ancient texts of Hindu tradition. Vedic yoga is closely associated with the performance of rituals and the recitation of mantras by Brahmin priests. The great contemplative dialogues recorded in the Upanishads are distinguished by introspection and intense inquiry into the nature of the Self. These intimate dialogues between sages and their students distill the essential meaning of the Vedas: they express a movement from exoteric ritual to meditation on the great principle of unity that resides in the cave of the heart.

The ancient teachings of the Upanishads emphasize the path of jñana or knowledge of the Self and are best suited for persons who retire from the world and devote themselves to contemplation. A new phase of yoga emerged with the teachings contained in the Bhagavad Gita around the beginning of the Common Era. While the Gita endorses the path of wisdom, it stresses that one need not renounce the world. According to this text, yoga is skill in action. Krishna teaches Arjuna that he must act, but not be attached to results or outcomes. Their dialogue takes place on the eve of a great epic battle. While the battlefield is an effective dramatic backdrop, the Gita's subject is the war within, the struggle for self-mastery that every person must wage if he or she is to live life wisely and skillfully. The Gita thus offers approaches to yoga for different kinds of people: contemplatives, householders, and others fully engaged with the world. Krishna also teaches the path of devotional love (bhakti) and reveals this to be the secret or essence of the soul's relationship to the divine.

The Gita progressively integrates existing religious and philosophical systems into an encompassing world view, a synthesis that reinterprets and absorbs aspects of Vedic tradition, teachings from the Upanishads concerning the Self, the philosophy of Sankhya, and the classical yoga of Patanjali. The main thematic structure of the Gita is a description of four yogas – knowledge (jñana), action (karma), meditation (dhyana), and devotional love (bhakti) – by means of which Self-realization and liberation can be attained. All of these strands intersect and are further refined in the Gita's expositions. The text thus expresses the great recombinant power of yoga and offers an inclusive path: one available to all irrespective of birth, race, gender, or caste.

The Gita, however, is not the only synthesizing perspective on yoga presented by Hindu tradition. An even more encompassing vision of life comes in the form of Tantra taught by the great yoga masters who lived between the eighth and twelfth centuries. The tantric traditions integrate the principles and practices of the aforementioned yogas, establish lineages of yoga masters who initiate their disciples into tantric forms of sadhana or spiritual practice, and provide new texts such as the Shiva Sutras and Spanda Karikas.

Tantra derives from the verbal root tan, meaning "to expand," and is also often translated as a "loom" to convey the idea that reality is a seamless whole that comprises transcendence and immanence, being and becoming, eternity and time, spirit and matter. One of the most important tantric movements is the non-dual Shaiva schools of Kashmir known collectively as Kashmir Shaivism. The texts and teachings of this tradition emphasize the importance of the Guru, the creative energy principle described as Shakti, the initiation transmitted by the Guru called shaktipat, and various modes of sadhana or spiritual practice. Analogous teachings and practices are transmitted through the Vajrayana, the tantric tradition of Mahayana Buddhism.

In both traditions, Shaiva and Vajrayana, the teachings and practices are transmitted by lineages of siddha gurus. This is the tradition of Mark Griffin and his Guru lineage, the modern source of which is Bhagavan Nityananda of Ganeshpuri. Swami Muktananda received shaktipat from him in 1947 and was empowered by him to initiate others, first in India and later in the West. Griffin continues the work of Muktananda by offering shaktipat initiation to those who seek it. In this book, he offers rare teachings on shaktipat so that those who receive it will be better able to understand and align with the psycho-spiritual process of awakening that unfolds in their lives.

The tantric teachings of Kashmir Shaivism are expressed in several complementary schools. The Pratyabhijna ("recognition"), a wisdom school, teaches that the goal of yoga is to recognize that one is Shiva. Shiva, in this context, means ultimate Consciousness or the Self. The Spanda school focuses on the divine feminine energy known as Shakti and teaches that the universe consists of vibrating awareness.

The great tenth-century philosopher-sage Abhinavagupta created a synthesis of these traditions known as the Trika ("the three") system. The Trika contains principles expressed in triads, the most significant being the Shiva-jiva-sadhana. The world is seen as a manifestation of Shiva or Consciousness and the individual or jiva is nothing but a contracted form of Shiva. The movement of the universe has two aspects: the arc of descent in which Shiva becomes a person and the arc of ascent in which a person becomes Shiva. Put differently, Shiva becomes jiva, the individual soul, via a process of contraction; spiritual awakening reverses the process: jiva becomes Shiva. Yoga sadhana or spiritual practice is the means for attaining the goal of Self-realization or Self-recognition as an embodied experience of liberation in the world (jivanmukti).

How is it possible for a spiritual seeker to attain such a lofty destiny? What kind of guide is needed to undertake such a journey? What energy is capable of transforming a bound individual into an awakened being? The answer given by texts such as the Shiva Sutras and the Guru Gita is that the Guru is the indispensable means. It is important to realize that the Guru is primarily a principle rather than a person. The Guru principle is the grace-bestowing power of God.

Griffin explains that the Guru is an eternal spirit, a foundational element in creation itself, as well as the inmost essence of a person. The Guru also appears through the agency of yoga masters and siddhas who initiate and guide the awakening process of others. The Guru is a messenger, the one who provides a wake up call to those who have forgotten their true nature. Our lives are characterized by forgetfulness or ignorance. We are enthralled by our myriad sense experiences and the spell of the world described in Indian tradition as maya. Ignorance propels the wheel of samsara, our repeated rebirths into embodied existence in a bound condition. The Guru is our means of waking up from this cycle of delusion. But how does this occur?

Griffin details the answer to this crucial question in his thorough description of shaktipat. In the Indian yogic traditions, shaktipat refers to a form of spiritual initiation that grants immediate entry into spiritual life. The term literally means "the descent of the shakti," the descent of the power of ultimate Consciousness itself.

The Guru is the agent who bestows shaktipat by transmitting spiritual power to the initiate. Through this transmission, the latent spiritual energy within a person, kundalini shakti, is activated and the root of spiritual ignorance is destroyed. Until then the dormant energy lies coiled in the muladhara or root chakra of the subtle body located in the region of the base of the spine. The energy is contracted or limited; when activated, the energy uncoils or expands, gradually ascending through the chakras until it unites with Shiva in the sahasrara or crown chakra. Shaktipat ignites this auspicious process and initiates a great journey that culminates in spiritual awakening. The initiate discovers that sadhana begins to unfold naturally as he or she cultivates the impulse of grace received through shaktipat.

The sadhana that Griffin teaches involves esoteric components like meditation on the mantra that spontaneously arises with the breath (so'ham), the subtle architecture of the human form, the pathway of the Guru (Guru margena), the awakened heart-mind (bodhichitta), the stopping of the mind (pratyahara), and states of meditative absorption (samadhi).

Griffin describes these dynamics and processes in detail in discourses that read like the oral teachings he gives to students at his weekly meetings, meditation intensives, and retreats. His teaching method, in fact, belongs to the tradition of "pointing out instructions," wherein the teacher points out the enlightened qualities that already reside within his students.

As a religious studies scholar who teaches Hindu and Buddhist Traditions, I have studied the concepts discussed in this book for nearly four decades. By engaging a yoga master who teaches that which he himself has realized through direct experience, my understanding has been immeasurably deepened and enlivened. Griffin, in my view, is an authentic Guru who freely and generously transmits the teachings of his Guru lineage, the great siddhas of Ganeshpuri, Swami Muktananda and Bhagavan Nityananda.

The teachings in this book present an integral yoga, one that incorporates what is essential from the historical traditions mentioned above. Griffin offers methods for quieting the mind, introspection and self-inquiry, focusing intention, cultivating love, and acting in alignment with divine will. In this way, he teaches a sadhana that bestows both bhukti and mukti, worldly fulfillment and liberation. Such is the aim of the Yoga Tantra, the path of the siddhas or accomplished ones.

The good news conveyed in this book is that spiritual awakening is possible; everything we need is contained within the human form. The obstacles that occlude our understanding may be removed and the splendor of the Self is waiting to be revealed. While there are many paths to awakening, few are as powerful and swift as the

Guru yoga described in this book. Shaktipat is at the very heart of that path, a transmission of grace that spontaneously kindles our hearts and illumines our minds. In these pages Mark Griffin points the way.

May all who read this book and practice its teachings recognize the Self and express this realization for the benefit of all beings.

Dr. Patrick Mahaffey, Chair

Mythological Studies Program
Pacifica Graduate Institute
Carpinteria, CA

---

[1] The Yoga Tradition: Its History, Literature, Philosophy and Practice (Prescott, AZ: Hohm Press, 1998), p. 7.

ॐ

CHAPTER ONE – WHAT

My Guru was Swami Muktananda Paramahamsa. Muktananda was a Siddha and thus his method of teaching was one of direct experience. His teaching emerged from the idea of generating the opportunity for a person to step into their own awakened condition, which lies spontaneously, arising within you, waiting only to be realized and recognized.

This idea is the very essence of mysticism – mysticism being an approach of direct experience. It's not one of theory or abstraction, but rather one of gathering oneself up and stepping directly into a condition of recognition of awakening. The ground to be cultivated and awakened is one's own life, one's own form, one's own being. Therefore all the parts of one's self are involved in the awakening process. This is the entire idea of the mantra SoHam.

SoHam is a primordial vibration, a mantra, a sound, but it is different from other mantras. It is said to be ajapa japa. In other words, it is a mantra that spontaneously arises, a sound that spontaneously emerges and it is said to be particular to the human form. In other words, when consciousness moves freely and openly through the human form, the vibration, the throb of the SoHam, spontaneously arises.

SoHam does not arise as one would repeat a mantra in one's mind or in one's mental voice or repeat it vocally, but the very movement of consciousness through the human form produces the throb of the SoHam, very much like a wind moving through a stand of trees produces a very powerful whooshing sound. It is a combination of form and consciousness that spontaneously emerges.

The Four Bodies

In the Yoga Tantra, the conception of the human form is not one of a single formation but one of increasingly subtle assemblies of consciousness. There are understood to be four essential bodies.

The first and outermost body is the physical body – one of matter, bone, flesh.

The second body, the subtle body, is the body wherein the life force, the prana shakti, moves. It is in the subtle body that there are three streams that move from the base of the spine to the crown of the head. The left and right channel are female and male energies. Ida on the left is female and red in color and pingala on the right is male and white in color. Through the center is the central nerve, known as sushumna.

It is on these three streams, these three rivers, that the six chakras, the six plexuses - assemblies of consciousness, appear. In the second body, the subtle body, there is one at the base of the spine, one at the pelvis in the region of the genitals, one at the navel, at the heart, the throat and the forehead. Each of these matrixes, wheels or chakras, has subtle streams of energy that roll off of them called nadis. There are 50 in number – four at the base of the spine, six at the second chakra in the pelvis, ten at the navel, twelve at the heart, sixteen at the throat and two at the forehead.

The lower five chakras assemble along the energies of elemental force - the earth element at the base of the spine, water element at the second chakra, fire element at the navel, air at the heart and ether at the throat.

The forehead is the terminal point where the red and white energies rise up from the base of the spine, go over the crown of the head, come down over the forehead and go down over the bridge of the nose. This is the ajña chakra, the seat of the third eye. It is also called bindu and it has an echo at the back of the head.

At the very crown of the head is a maha chakra that is a thousand petals, a thousand fibers of light, and they bundle at the crown of the head and flow upward into the space at the crown of the head. There are other streams of force – 17 of primary number. Of the 17, the three that are the most important are the ida, pingala and sushumna, and then there are lesser and lesser, up to some 72,000, known as nadis.

It is inside these 72,000 nadis that the karmic assembly of the cause and effect of one's actions are stored.

The third body is the causal body, the mental body, or the body of mental formation, the assembly of "I". There are many subtle planes of consciousness where awareness operates so uniquely that they can be identified by their patterns of behavior and vibration.

The three most familiar aspects of the mental body are 1) the ego reflex of "I" consciousness, 2) the list-making capacity that orders the operation of the senses and supports the separate identity of the "I" consciousness and 3) memory, which also produces the definition of the unique nature of separate identity.

Whenever someone asks who we are and how we are, the first thing we tend to do is make a reference to our stream of experiences because we are shaped by them and our identity is drawn from the compressed assembly of all those experiences that we call memory. These memories and experiences have an analog in the subtle body in the form of stored energy called samskara. They're like bits and bytes of data and they're stored in the 72,000 streams of the nadi. So there's a correlation between mental formation and the pranic life force, which is in the second body.

It is this entire conceptual collection of samskara that we call karma. Karma is the weights and balance of cause and effect. When you hear of spiritual practice generating purification, this is what is being purified.

That stored data in the subtle body also has a correlation in the physical body. We feel it as psycho-physical tension that is stored in the connecting tissues of the muscles, the skeletal system, etc. where there are powerful psycho-dynamic energies trapped in the tissues of the body. They're also stored in the subtle body as these bits of data called samskara. All together this is called karma.

The reason a person doesn't arise realized

and immediately recognize their inherent true spiritual nature is because of this body of karma – all these bits and bytes of samskara that are assembled in these countless channels of prana in the left and right channel and the 50 fibers, which actually double in each site. Each of the 50 fibers has a male and female corollary, so there's said to be 100 fibers that flow from the base of the spine and from each of the chakras, throughout the physical body and subtle body.

All the 100 fibers terminate in the brain. We know that we have six chakras and there are six basic lobes in the brain, so there is an immediate corollary to this structure that flows from the base of the spine to the crown of the head.

The ida, pingala and sushumna go up the back and hit a space at the base of the skull. When you see a picture of the brain, you see a series of fibers that is the white matter, gray matter and pink matter of the brain. Through the center of the brain is all white matter, right through to the third eye. The two basic glands of the pituitary and the pineal gland are right on that pathway.

There's a very powerful dynamic in the process

of the birth of a new body and the downloading of the new brain. The brain is actually empty when you are first born and all of your karma hasn't really downloaded. Anybody that's seen a newborn baby notices that they're just unmarked and unstained. There's a process over the course of months and years where you feel the baby's life force begin to attach to the body, download into the brain and you see the person start to emerge.

The fourth body is the supracausal body. The prefix 'supra' means before, 'causal' means consciousness. So supracausal means before consciousness. The fourth body has the form of a single scintillating blue atom. This body is the body of universal consciousness. It is very subtle. It is difficult to see in the unawakened condition because in the physical body we have the waking state, the sleep with dream state and the deep sleep state and we're always cycling between these three states. We're always moving from the waking state into the sleep with dream state, sleep with dream into the deep sleep state, then back out. Deep sleep state, sleep with dream state, waking state. That's the cycle you go through every night when you go to bed. You go from waking, sleep with dream, deep sleep, sleep with dream and waking.

All of these three states are attenuated to the conditioned body of the physical, subtle physical, and causal body. Those are the bodies that are in constant movement and constant transition. They are said to be transitory in nature.

One of the ideas behind the revelation of the fourth body, the fourth state, is that it is a state of consciousness that is constant and the same at all times. It is universal in its nature and it is the doorway to ever deeper consciousness. Think of it as the substrata behind the creation. It is the basis upon which everything appears: the envelope of physical matter; the envelope of the subtle physical, which is of the nature of electricity; and the envelope of mental formation, also known as the causal, which is of the fabric of awareness.

The Four States of Consciousness

It's important to have an idea of this picture because how you're put together has everything to do with how you awaken. When you are in the waking state you are ruled by the awareness of the body, the awareness of the senses, the operation of the mind and the movement of the life force. You're aware of being alive. When you move into the sleep with dream state, something very different happens. It's a very dramatic shift of consciousness. You're no longer aware of the body. You are no longer aware of the senses.

In the dream state you're only aware of the mind and the memory operation of the mind that is generated and activated in place by the life force. When you go into deep sleep, again there is a very profound shift of consciousness, in which the mind thinks of nothing at all. In REM sleep we see the activity of dream. There's lots of activity in the brain. As one dreams, you reproduce the data of the senses so you are aware of the memory of the senses, but the senses no longer operate.

There is something very unique about the difference between the waking state and the deep sleep state. The waking state is a shared state. We plug into the vent of it through the senses, and we agree upon its format.

But when we go into the sleep with dream state, that is not a shared state. Only the dreamer experiences the dream, yet the experience is as vivid if not more vivid than the waking state. It's extremely intense and can be electrifying or terrifying or charming based on the emotional data that's being released in the dream. It's very interesting because someone can be asleep in the room, and they can be surrounded by thirty or forty people in the waking state. Even if their bodies are in the same room, their awareness is operating in two different realms of reality. The person that's in the dream state will not have any idea of who or what was going on in the room around them. And the people sitting watching the person dream will not have any idea of the content of the dream. They can see the dreamer's body, but they can't see into the dream.

When we go into the deep sleep state the mind

goes into a state of neutrality. It stays active, but it thinks of nothing – no thing. This is the most restful of the states and it is where the entire system rejuvenates and reboots. If we go through a situation where we have a rough week and just sleep very lightly, we might only get into the sleep with dream state three or four nights in a row. By the third or fourth day we would start to feel kind of frayed or wired not having gotten into the deep sleep state. But then finally we'll get into the deep sleep state and wake up feeling very refreshed with a spontaneously positive outlook about life and the world.

Most people will live their entire lifetime going from one of these three states – waking, sleep with dream, deep sleep, sleep with dream, waking over and over again, not noticing that they are in a closed circuit. Every now and then they'll swing past a door that's on the deep inside structure of being, and that doorway is the fourth body. It's vibrating and present, and animates and pervades the three relative bodies and also the three relative states. But very few people have direct experience of it.

In their wisdom, the sages simply call it

the fourth state, which is the state of direct recognition of the universal nature of one's consciousness. It's present within you as you. It always has been and always will be, but it's most often unrecognized.

It goes unrecognized because of the constant nature of the phantasmagoria of the dazzling senses in the physical envelope, along with their operation in memory, and because we continuously cycle between the waking state to the sleep with dream state to the deep sleep state and back. It's such a drumbeat over and over again and what happens? You forget the fourth state exists.

So the sages simply say, "You're doing great as far as it goes, but you're forgetting one very important thing. You're forgetting the existence of the fourth state, and you're forgetting the existence of the fourth body, which is present and resonating within you as the Atman, the Blue Pearl."

They want you to slow yourself down and train yourself to directly recognize and experience the fourth state. That's it. Simple.

The Pervading Nature of Prana

The sages say the one thing that operates the same in all four bodies is the life force. When the life force is operating in the physical body, it's the breath. It operates inside the physical body in the form of the electrical impulses that run the voluntary and involuntary systems that animate the brain.

Ninety percent of our entire experience of the waking state is dominated by the senses. The prana shakti, the life force, goes into the subtle body where the envelope of the prana is the most dynamic and declares all its seats. It's still the prana, but it operates differently because now it's operating in the subtle body, which is life force that is moving up and down, riding in and out of the body on the breath, descending the in-breath, ascending on the out-breath. Every time the breath moves, the prana is riding as if the breath were a horse and the prana were a rider. Every time the breath moves, the prana moves.

And that gets us to the third body, the causal body. Prana shakti, the life force, is operating in the third body, the body of mind and mental formation. The three most familiar aspects of the mind are the structure of ego, the list making capacity of the intellect, and the storehouse of our memories – all of which are floating in a substance that we call mind essence, chitta.

Even though it's in the causal body, the mind rules how the prana operates. The prana is still the prana. It's still the life force. It's still the vitality – the vitality in the causal body of mental formation, the vitality of the prana, the life force in the electrical body of the subtle body, the vitality in the physical body animating the movement of the senses and the mind etc. It's always the prana.

Again you can see the wisdom of the sages saying, "Watch the breath." They tell you to watch the breath because they want you to notice that the vitality that gives you the experience of being alive is moving on the breath – in and out of the body. They want you to pay attention to that and to see it. It's a subtle vibration.

And that leads us up to the potential for the realization and recognition of the fourth state.

The life force emerges out of consciousness itself and appears as the body. It moves into the subtle body as the electrical impulse of the kundalini and the life force, and it engages the incredibly complex circuitry of the subtle body. It moves into the mind, which is formless in nature.

The mind is mirror-like, without a form. It reflects what it comes into contact with. Have you ever noticed when you're in a beautiful place, your mind goes to a beautiful place. When you are in a dark and scary place, your mind operates in a dark and scary way. When you're with a cheerful and positive person, your mind operates cheerfully. When you're with a negative person, your mind will tend to operate negatively. It tends to pick up on its reflections. You'll have the momentum of your own memory and tendencies, but the mind is very powerfully influenced by what it comes into contact with. That is why the sages always advise: Keep good company. Keep the company of the Holy Sangha – those fellow beings that are seeking enlightenment. You'll be carried along by their enthusiasm. Your enthusiasm will help them, and as you face the world you'll find a quality of inner-connectedness. You can never run from the world but you can seek powerful allies.

So that brings us to the idea that the prana operates the same in the first three bodies. Well guess what? It operates the same in the supracausal body – the fourth body, which merges back into consciousness. It's still the prana. It's vitality. The journey from consciousness to creation and back to consciousness is a subtle one, that is revealed in the breath. If you begin to understand how the human form operates and is put together, you can use this principle to awaken more and more to yourself.

Over time you come into direct experience of what the prana represents in relationship to the four bodies. It's like a diagram. You have a TV. You take it out of the box. Plug it in. Hit the on button. Look at the instructions. It's an entire diagram of everything in the television. Even after you read it, you don't understand it. There are only a few people that actually know how a TV works, but everybody watches it.

The human form is much the same way. You

have to pay attention to it to begin to understand it. And that's what meditation is about. It's simply about understanding the underlying workings of your own being.

It is the movement of the prana that is your chariot, your vehicle, that without effort spontaneously carries you through these various sheaths of physical, subtle physical, causal and supracausal. Even though they are layers of consciousness, they are not separate consciousnesses. They are the same, but consciousness operates uniquely in each of these four bodies. It's the ability to penetrate through the layers and finally begin to directly apprehend the operations of the fourth body that makes the process of meditation useful. And this is what meditation is about.

When you come to the experience of the fourth body, that is the experience of God. When you have any kind of contact with the fourth body directly, you feel you have come into contact with God. As you come closer and closer, that experience of the apprehension of God becomes overwhelming. Then there's a point where they blend and melt into each other and the individual identity is lost and absorbed by the universal identity and the circle is finally completed. And the purpose of spiritual practice has been fulfilled. A new basis of identity is established by closing the circuit between the waking state and the fourth state.

What is so interesting is that each of the states is so unique and separate from each other. It's almost as though each one takes place on a different planet. One is not even remotely like the other, yet they appear inside the stream of a single human being.

In this school of mysticism, the sages try to simplify the teaching because of the wild phantasmagoria of the operation of the senses and the intensity of the three relative states of waking, sleep with dream, and deep sleep. This is why it's very useful to have the understanding of the life force, the prana shakti, and the breath. This is where the SoHam comes in.

SoHam and Kundalini Awakening

The breath is the key to the meditation process. You walk into any meditation hall in any culture or tradition and you say, "I want to learn how to meditate". The first thing the meditation teacher will tell you to do is to sit down, still your mind, begin to breath deeply and watch your breath.

What they want you to notice is something that is very important. Not only is the breath just a movement of gases inside and outside of the body, but the vitality of the life force, what we call life itself, prana shakti, is attenuated to the breath.

The inward breath enters and descends, the outward breath rises and expands, and inside each breath is a vibration, a throb of pure vitality that is the prana shakti. This is the first face of an energy of infinite power that is called the kundalini – the great force of creation.

The SoHam operates on two great energies. It is the great descending energy and the great ascending energy. The SoHam comes from what the Yoga Tantra describes as this polarity. The entire basis of the tantra is founded on this equilibrium of polarity. And the pathway of equilibrium are these three rivers that go from the base of the spine to the crown of the head. This three foot structure is the spiritual path. The spiritual path is not outside you. It is the illumination of every inch of that three foot pathway within you from the base of the spine, up through the brain.

The SoHam is the Shiva Shakti principle, or the yin and yang principle, female and male principle. The great ocean of consciousness which is said to be Shiva, and the infinite manifestation of creation which is said to be Shakti. They are not separate. They should be conceived of as the sun and the rays of the sun. There's no point where you can say, at this point Shiva is no longer valid and Shakti exists, or Shakti is not present and only Shiva exists. They thoroughly pervade each other. You have to understand this idea because it is a paradox to think of two things as one thing just because it has two names. Normally if something is two words, then it must be two things, but it's not.

There is an appearance of separation based on how perception occurs.

The ancient way of awakening is one called shaktipat in which the kundalini is generated and moved from a dormant condition into an active condition. This causes the life force in the kundalini to go from its potential condition into a dynamic condition. That means that every time the breath moves in and out of the body, the dynamic force of the life force acts like an ever-increasing purifying flame and it removes a little bit more of that samskaric and karmic data that's stored as information inside the subtle body.

This karmic data has a physical presence. It has a gravity. You can weigh it. And it can be eliminated by bringing it into contact with the essence of the creation itself, which is the prana and the kundalini. With each breath we have an extraordinary inward flowing force of pure vitality.

What changes its nature? It's quantum mechanics. Quantum mechanics is the idea that the perception of any object is changed by the very perception itself. Perception changes the nature of what is perceived. That principle operates very deeply in the spiritual life.

As in quantum physics, the fact that the breath is conscious increases the power of that breath. The fact that you're aware of the prana present in the breath animates it and changes its nature, causing it to purify the system.

The great descending and ascending forces are expressions of Shiva and Shakti that you see over and over again in the universe. It's always these dynamic polarities that are seeking an equilibrium.

You see the left and right channel seeking equilibrium in the central channel. In the human form when the consciousness is moving in the left and right channel, consciousness is externalized. Through the power of meditation you can move it into the central channel where the mind withdraws from the external universe and merges into the Self.

The kundalini exists at the base of the spine in the form of a coiled serpent and indeed that is the kundalini in its most essential form. Kundalini means coiled serpent. When you generate the movement of the breath, the movement of the prana, the kundalini vibrates.

The kundalini stored at the base of the spine is associated with the syllable So – the syllable of the creation. The vibration of So gives rise to the appearance of the creation – infinite universes, the worlds, and all of the relative assemblies of the five elements, earth, water, fire, air and ether. The crown of the head vibrates with the syllable Ham. Ham is the vibration of Shiva, which represents the ocean of consciousness.

We oftentimes hear about the kundalini rising up through the central nerve and the three rivers to merge with Shiva at the crown of the head. We see this incredible structure with the six seats going up to the forehead and then over the head there's this huge wheel with a thousand petals, called the sahasrar. That thousand petals is a bundle of fibers that goes up through the spine, through the center of the brain and merges out of the crown of the head at the brahmarandhra – the soft spot at the crown of the head, and continues to go up into the space over the crown of the head.

We've been talking about the great descending and the great ascending force, and we see these energies in all creation. We see it also very powerfully in the human form and in every breath. When we breathe in, the breath goes in and down; it sinks. On the incoming breath, the Ham syllable at the crown of the head rides down through the center of the body, through the brain, through the throat, heart, belly, loins, into the base of the spine, and strikes the So syllable at the base of the spine.

On the physical side the breath descends. Remember that the prana is the same in all four bodies. The prana is riding inside the breath as a subtle throb, as a vibration. That is the Ham vibration moving down through the three rivers striking each of the plexuses, each of the chakras, animating them and as it strikes them, it purifies and radiates out through the hundred fibers that go throughout the entire body and all terminate in the brain. As it hits the brain it purifies and awakens the brain mass.

As it strikes the base of the spine the So syllable

is stimulated and the outward breath begins. And when we breathe out, it rises and expands. And thus the So syllable at the base of the spine and the Ham syllable at the crown of the head change place with each cycle.

Every time Muktananda would talk about SoHam he would always move his hands up and down to describe it. It's important to remember that it's not sequential. It's incorrect to think that it's only moving down or it's only moving up at any one time. It actually moves up and down simultaneously. What's moving is your attention. Your attention is jumping from energy to energy.

The vibration of the SoHam is a great descending and ascending force that is of the nature of the ocean of consciousness and the ocean of universes of the creation simultaneously. What is the pathway? The three nerves ida, pingala and sushumna that go through the center of the body. What is the vessel? The vase of the human form – physical body, subtle body, causal body and supracausal body. With each movement, with each throb, a little bit more of the content of the stored memory and cause and effect is removed. And that effect is like one of clouds being burned away by the sun revealing the true sky.

This kind of meditation is cumulative. You take twenty-one thousand breaths every day, but how many of those twenty-one thousand are taken consciously where you are vividly aware of the vibration of the prana, the vibration of the So and Ham in the descending and ascending breath? Maybe just a few, if any. This is just fundamental metaphysics of how consciousness is moving inside you. It happens one of two ways. It happens consciously or it happens unconsciously. The sages say: "Make it happen consciously. You will see something. You will experience something. You will realize something true about yourself."

Imagine just for one day you took twenty-one thousand fully conscious breaths. By the end of that day you would be a completely transformed person. You would not be even remotely the person that began that day. If the next ten breaths you take are fully conscious you'll be completely different after ten breaths because this kind of breath acts like an opening and awakening. It goes into the innermost subtle spaces of being. This exact same impulse of energy causes a bud that is tightly bound to open as a flower. You can't tear it open from the outside. It's just a mess. You have to send that impulse in and then it opens. This is what that's about. It's simple. It's a fruit in your hand. If you understand it correctly then this process opens the physical body, the subtle body, and all of the plexuses and the various subtle streams of each of the chakras.

Each of the petals of each of the chakras will all go from a bound and closed condition into an awakened and illuminated condition. As they begin to illuminate, that inner system begins to be infused with light, which is where the word illumination comes from. You begin to glow from the inside out.

Each of the four bodies experiences the process of illumination through the ever expanding awakening of the kundalini, the inner vitality. As they begin to pick up momentum, the energy between the crown of the head and the base of the spine begins to take on the nature of a thunderbolt, like a constant flow of electrical energy. The Zen masters say, "If you can maintain the awakening breath with the constancy that a slate tile dropped at the surface of the ocean descends towards the depths of the ocean without a single break, you'll be completely enlightened in seven days."

Meditation Instructions

The best way to breathe, especially when you're doing a meditation session, is to do a form of breathing called the Bellow's Breath. It is called that because you treat your body and subtle body as a bellows. When you want to charge a bellows, you pull it open. When you want to discharge it, you compress it. So thus when you want to breathe in, you expand the diaphragm, which opens the entire upper body and the breath comes rushing in. When you want to breathe out, squeeze the diaphragm, just like discharging a bellows, which pushes the breath up and out.

Take a minute to put your hand on your diaphragm. Breathe in by pushing the diaphragm out. Wait a beat and then breathe out by squeezing the diaphragm. Just take a few breaths like that. Note how it feels. You should feel a very powerful descending energy striking straight down through the center of the body from the crown of the head down through the heart, down into the belly, through the pelvis, to the base of the spine. You'll feel it as it strikes each of these chakras. Because the chakras are like centers of magnetism, you'll feel the kundalini pool and move there. It's also useful just to breathe directly into the heart, maybe every third or fourth breath, because it's exactly in the middle – three higher chakras and three lower chakras. It's a very good seat of equilibrium as your meditation settles ever more deeply into your system.

Another key point for deep meditation is to be aware of the space between the breaths. As you take a breath in, notice that you're breathing in, you're breathing in, you're breathing in. Then there's a point where you are no longer breathing in, but you are not yet breathing out. Then you're breathing out, you're breathing out, you're breathing out. Then there's a point where you are no longer breathing out, but you are not yet breathing in. Put your focus on these spaces in between each in-breath and each out-breath.

When the breath moves the prana moves. In between the breaths there's a point where nothing is happening. The breath is not moving. The breath is still. Because the breath is still, the prana is still. Because the prana is still, the mind is still. And at that point it's an opportunity to move from the left and right channel into the central nerve, into the sushumna. It will happen naturally.

The main point I want to make is to pay attention to the breath, going in as it descends, going out as it ascends and also in the space between the breaths where it is neither falling or rising. Press your attention into that place. Something very important is happening there.

Listen to the sound of the breath. You don't have to repeat the mantra. As you become subtly merged with it, you'll see that the mantra is arising from a very deep place. It is said to be ajapa japa. It is the mantra that repeats itself. It is the essence of the Siddha path. It generates spontaneous awakening of the totality of the human form and finally of the universal form, the gateway of which is the fourth body.

With each breath and the impact of the SoHam, your being looks very much like an opening flower. As that occurs you'll see more and more of the inner content of your own nature at these various levels of consciousness. Be very fearless and attentive. Watch it as it unfolds. You'll notice that the impact of the SoHam meditation is cumulative. It will gain force with each cycle.

If you try to hold it all in the physical body, you'll just get this sense of incredible pressure, and you'll become physically and mentally uncomfortable. But if you allow the breath to open and relax, particularly in the second body, you will move into deep meditation easily. So employ the bellows breath.

It's very useful to stay physically and mentally relaxed. Make gravity your friend because the energy of the kundalini has a strong electrical component. It's important to keep the spine straight because the pathway of the energy is from the crown of the head through the brain down through the spine.

CHAPTER TWO – WHY

The Guru and the Guru shakti are a great mystery. Suffice it to say in the simplest terms, the Guru is the grace bestowing power of God. It is said that the Guru is the incarnation of mercy and the essence of awakening.

The essence of the Guru is an eternal spirit that arises at the inception of creation. It is a part of the creation and is not separate from the ocean of consciousness that is the origin and cause of all things.

The expression of these two words, Guru and shakti, carries a very powerful message. The word Guru represents the grace bestowing power of God; the word shakti translates as power or force. It actually translates as the power of whatever word it is connected to, so Guru shakti refers to 'the power of the Guru'.

In the spiritual hierarchy, the shakti represents the creative force – what we call the kundalini in the yoga tantra, a force of infinite consciousness, infinite quality beyond all conception, out of which everything is assembled and everything is made. The word "Guru" means the awakening principle of consciousness, which stands as a bridge between light and darkness, between everything and nothing and between the ocean of consciousness and the manifestation of the creation.

This profound polarity is found in all philosophical conceptions of the origin of creation and the reflex of cause that arises unseen – yet manifests. The principle of the Guru is that which ripens, that which fulfills, that which completes. All things come into creation and go through a kind of cycle.

Samsara, The Wheel of Cyclic Existence

Although we manifest completely as the expression of the creation, the totality of the creative impulse and the ocean of consciousness is fully expressed within us. In the passing between the two layers of consciousness there is a kind of frequency of forgetfulness and a loss of knowledge of Self.

This forgetfulness arises as a subtle form of ignorance. In Eastern philosophy it is referred to as maya. Maya functions by generating a consciousness of existence. The state of 'I am' and the energy between these two polarities produces a cycle of self-awareness and the creation of an 'I'. In the creation of an 'I', a polarity of recognition is set in motion as a condition of contact.

We establish a field of consciousness that would be characterized as fields of perception. It is through these limited fields of perception that we come into contact with the creation. It is consciousness seeking to perceive itself as it comes into contact and generates sensation.

As it generates sensation there is a quality of attachment.

When you watch how your mind operates, there is a subtle idea that one thought follows another. When something happens once there is an attachment to its manifestation, and we seek to cause it to happen again. In this expression of attachment there is a development of grasping.

This grasping produces a subtle bondage to perception itself. Enormous spiritual resources of consciousness, energy and life force are expended in this process. This grasping produces an energy that we commonly call desire. It is this desire that is the root source generating birth, life, decay and death. This is the cycle of the operation of consciousness as a single perceiver.

In the spiritual philosophy of yoga tantra, this cycle is called the wheel of cyclic existence, or the wheel of samsara. You will find that almost every perception that you have will fall into one of these categories of experience. This cycle has a momentum to it, like the principle of momentum in physics. When something is set in motion it tends to stay in motion.

These energies feed into one another producing a bound condition. It's as if you were inside an ocean of infinite consciousness but the momentum of your perception, your individual consciousness, is moving around the spokes of this wheel of separate origination with such force that you actually quite literally lose contact with the ocean in which you are suspended and through which you are traveling. The ocean of consciousness is flowing through your every cell, your every atom and flooding your every breath.

The Guru Principle and Shaktipat

With this picture established, it is the Guru and the Guru shakti that provide the salvation and comes up to your individual bubble and says *"Knock knock, is anybody home?"*.

This is shaktipat, and it is quite literally done this simply. It is basically a contact. It's just a simple touch. That's all it takes. You can be touched by hand. You can be touched by line of sight. You can be touched by thought. Once the Guru touches you, this force of awakening begins to consume and transmute every one of your atoms and every one of your particles of being. The entire force of creation is behind it.

The effect of shaktipat is that your attention is redirected to the ocean of consciousness in which you are suspended, which you have quite literally forgotten about in the momentum and thrall of the maya that's taken place as you move around this endless wheel of cyclic existence.

Picture in your mind that you are a single drop floating in the endless ocean and you have forgotten completely about the ocean. You are happily living inside the drop. The Guru principle is that force that comes up and goes *"Tap tap tap! Behold the ocean"*. As soon as you turn your consciousness to apprehend it, it comes pouring in and the drop and the ocean begin to merge, becoming a single thing.

This is why the Sufis say the ocean is in the drop and the drop is in the ocean. It goes both ways. This is shaktipat. It's not some wild idea that took place at the edge of time and is a leftover from a previous era. It is a principle that is constantly present and always active, and goes through different phases of appearance and nonappearance.

In all philosophy you will find references to the existence of this absolute truth that pervades you completely. These are the references to what we call God and are present in all the social structures that have institutionalized the God concept. And in so institutionalizing them have almost guaranteed that you will never turn your head and look out the window and see what's going on. Religion produces a kind of institutionalized forgetfulness. It's where God and forgetfulness get together and become something truly difficult to overcome.

The Guru has a very simple approach. The Guru is this principle of awakening. This is why the Guru is characterized as the incarnation of mercy. The Guru principle arose at the point of creation and as an integral component of the creation. It is simply a wake up call. It is the Guru that drives this ripening energy through every life form, however simple, however complex, however foreign, however familiar.

The Guru shakti is the point where this inconceivable amount of power and spiritual force comes to a point of awakening, where we become aware of our dream taking place inside the dream of the creation. What the Guru does is ring the bell that wakes you up and breaks you out of the thrall of this infinite momentum of the wheel of cyclic existence. The formula of the spokes of this wheel is so strong that it produces a tone of forgetfulness of the substance out of which all consciousness emerges. It is the Guru who brings the fabric of consciousness back into the formula.

The Guru is an eternal spirit that arose at the

moment of creation, is present when the creation fails and is present again when the creation comes back into existence. The Guru is that part of being that is luminous and self-aware.

It's very important to understand this. I'm speaking in vast absolutes, beyond comprehension. Yet at the same time, this incredibly vast force applies itself to you, through you personally. The Guru shakti is eternal. We can think of it like a ray of consciousness. If the infinite creation is countless rays of consciousness, then one of these rays is the Guru shakti. It's a radiance that is present in every molecule, atom and particle. It is the chit itself.

Image this vast consciousness as a kind of stew or soup, where some parts of the stew have already been touched by the Guru shakti and that awakening impulse begins in that little atom. But there are other atoms where that hasn't happened yet. So there are awakened and unawakened particles of being all stirring around together, just as there are awakened and unawakened sentient beings. There are beings that have not been touched yet by the Guru principle and other beings that have been.

You will sometimes see that the Guru shakti will begin to saturate a particular being. And perhaps this being has been in the flow of origin and the flow of existence for countless ages, so much so that their consciousness is pervaded by the consciousness principle that we call the Guru. These beings do exist. There's no part of their being that is unawakened. There is no part of their being that is dark. They manifest in 10,000 world systems simultaneously because there's no limit to their consciousness. Why? Because they have so aligned with the Guru shakti that there's no part of them that is resistant to that presence. There's no part of them that remains in darkness. We hear of these beings down through history and their stories appear again and again, pervading all cultures and systems of belief and religion.

The Guru principle manifests as a particular operation of consciousness in a given time and place. The effect of the Guru shakti that is present in a given time and place will have an extraordinary impact on the sentient beings present at that time.

The Guru shakti is a very special impulse that is almost impossible to describe. It's like an algorithm that processes or transmutes the inconceivable into food, energy and spiritual essence. The inconceivable is the kundalini, the force of creation, which is like the sun. Consider the sun for just a moment. It is well understood that life on earth is dependent upon the balance of elements present here on this planet in relationship to the distance of the rays of the sun – that incredible force of light and heat. Seven-tenths of the planet is being given over to water to modify that incredible force. If it were just a few thousand miles closer it would be a burning ember; a few thousand miles farther away it would be an iceberg. In other words, it's an extraordinary balancing act.

The kundalini is an ocean of infinite and inconceivable force, and it is expressed in every molecule of not only this universe but endless universes. It's also expressed in every molecule inside you, as you. You could never come to the end of it, and if you come even somewhat close to it, you could be burned to ashes in a second – your mind and body destroyed. If the kundalini is the sun, then the Guru is this principle of balance.

The Guru is the principle of awakening and remembrance that rings the bell and says *"Hello, this is what you've forgotten"*. At the same time it directs your consciousness into that inconceivable force of creation that is sun-like, even though saying that it is sun-like is an understatement.

But at the same time, the Guru modifies this energy of the kundalini. It's just like the principle of a transformer in our electrical system. The universe is basically electricity. By modifying the kundalini, it transforms this energy into a useful vibration. So rather than being incinerated in a moment's time, rather than being obliterated by the energy of those billion suns, that energy is made gentle. It flows into you and through you and atom by atom, particle by particle, vibration by vibration, it begins to awaken you.

Your consciousness begins to break out of the shell of the thrall of the maya that has been held in place through the momentum of the wheel of cyclic existence. By the virtue of the power of maya, you've forgotten the true nature of yourself, the true nature of your origin. Suffice it to say, this wake up call is very powerful.

By the time you've gone through countless lifetimes, thousands and tens of thousands of incarnations, your consciousness has been absorbed by the momentum of desire that is expressed in the basic spokes of the wheel of samsara. It is a very powerful, vast and totally complete shift of paradigm.

This is the way the Guru shakti works. It does not take recourse in a system of beliefs. It absolutely does not create a new religion. It simply breathes the life of truth into you and you are transformed. This takes place from the inside out. The quality of this touch, this knock on the door, this wake up call, is called shaktipat.

The word shaktipat translates very beautifully as the descent of grace. Grace is a gift. It's not something you've earned. It's a gift and it's a form of favor. It's the moment the sun is shining in your backyard that day.

The Guru arises as the expression of the mercy of God, and is the impulse of awakening and mercy, eternal and endless. You should understand that the Guru and God are the same. The totality of God is present in the operation of the Guru. In the situation where the Guru shakti is being embodied by a sentient being, what you're often times dealing with is a being that has been existing so long in the cycle that they have been saturated by the Guru shakti over countless lifetimes, and so they've become an agency of the Guru principle.

Some beings that manifest as the Guru shakti were never human. They're simply a manifestation of consciousness that appears in order to generate shaktipat at a given time and in a given place. In other cases, perhaps the human beings started in the human cycle – started in the incarnation cycle maybe here in earth, came up through the trenches, manifested, got into the human form, received shaktipat and transformed and awakened. They may continue to exist as a form of service – because if the Gurus are anything, they are the servants of God, and they're all up and down the ladder.

One of the things that I've actually found and enjoyed all my life is that when you speak the truth out loud it sounds like science fiction, and I've always enjoyed that. It sounds so outrageous that it couldn't possibly be true, yet it is.

I want you to take a moment to appreciate what I'm telling you. In whatever form the Guru shakti comes to you and makes that contact – perhaps it happened while you were in the womb, perhaps it happens in the last moment of the last breath of a given life, perhaps you come into psychic contact with the virtual army of beings that are operating in agency with the Guru – this moment of shaktipat is something complete and total.

When you are given shaktipat, you are not becoming involved with the personal power of a given person or a given place. Understand again, the Guru is the grace bestowing power of God. The incredible force of that inconceivable unity is – for that moment, at that time – thinking about you. It is of the nature of awakening. It is of the nature of the elimination of ignorance.

All of the fluctuations of consciousness that are the fruit of maya-based activity and behavior as you move relentlessly on the wheel of samsara, are transformed. Your memories of being an individual identity are true, but they're like a dream.

When you think about the dream you had where you were on a road and you saw a tiger and then you rode the tiger down the road and you went up into the mountains and you flew in the sky – all of those are memories of a dream.

All of the activity of the wheel: the manifestation of birth and life, decay and death, the generation of the sensory fields, the expression of reaching out and touching the created universe, of sensation and contact, attachment, yearning and grasping – they all happened, but they happened in a dream. They happened in the dream of cyclic existence. They're all transitory and as real as the dream you had last night.

When you're in the dream, it's completely real. Then the dream is over and another dream starts. And when that dream starts, that dream is completely real. And as soon as you start that dream, what do you do? You forget about the dream you were just in. Now you're in this dream. When that dream comes to an end, you start another dream.

These dreams exist as a force of memory and are the glue of your identity in any given time. When the Guru shakti touches you and shaktipat begins, the energy of these dreams starts to become transformed into the light of pure consciousness. In other words, it's a paradigm shift, a foundation shift. The foundation of your identity is being changed.

Once you've been touched by the Guru, once shaktipat has occurred, this is what happens. It can't be stopped, it can't be resisted and it cannot fail. The force of awakening is very powerful and absolutely irresistible. It begins to move through you and change you.

Once shaktipat occurs it's not so much that there are really any particular practices that you must do. The practices are there to help you deal with what's already going to happen – practices such as the eightfold yoga, pranayama and breath control, the mental control of concentration, meditation and the ability to stop the mind, and most importantly, the ability to enter into samadhi, where the mind can move into the state of consciousness in which it actively includes the existence of the ocean of consciousness and the infinite force of the kundalini.

Spiritual practices are a way to interact with a very profound process. If you were to take the practices of meditation, or the techniques expressed in the Patanjali Yoga Sutras or the Shiva Sutras, they would have x amount of power and benefit if you performed these practices without having a Guru and had not received shaktipat. In other words, the practices themselves would only have a limited impact on you. But once you receive shaktipat, everything changes. After receiving shaktipat, all the spiritual practices are tremendously amplified. By understanding how the process of awakening is conducted inside your system, you can move into an alignment of cooperation and inter-connectedness with a process that has already been set in motion by shaktipat.

Practices such as meditation, the use of mantra, the understanding of how the prana interacts with the body, and so on, are all useful. They're what an exercise regimen is to an athlete – they improve your skills. The talent is present, but the harder you work, the better the results are.

I've been involved in spiritual training my whole adult life and I've met all kinds of beings and many very adept and powerful yogis, and their level of practice was quite literally astounding. I received shaktipat when I was 21 years old and I'd already been using the word guru in a sentence for some years. I was looking for a guru and I was fortunate to find a really hot one. When I got shaktipat and started the practices, chanting mantras and meditating as much as I could stand, I started having extraordinary experiences, yet I found that people that had been practicing for 10, 15, 20 years weren't even having a fraction of the level of experiences that I was having. I could tell that the difference was shaktipat, that the Guru principle had cracked my egg and opened my consciousness into the ocean.

My experience personally was always one of barely hanging on by my fingernails as this force came through me and changed my nature. It was like waking up from a long dream. That's the end result of shaktipat. You can start to sense and experience the absolute inter-connectedness of all things. It's quite impossible to describe, but consciousness goes through a very powerful transformation. Shaktipat is essentially the arrival of the ocean at your doorstep.

As the process of awakening occurs in a person, the dream of the previous identity has a momentum to it and remains in place, struggling to hold its position, even as it is being driven by this irresistible presence of shaktipat. Eventually that person will find that certain qualities of life and mental attitudes begin to lose their thrall and simply fall away.

Sometimes there will be a phase of confusion between the old way and the new birth that is taking place spontaneously with every breath. This is where the use of spiritual practices comes into play because they are very powerful methodologies that tie you into the incredibly complex architecture that is a human being.

Remember, the human form is composed of four bodies – physical, subtle, causal and supracausal. Each level of consciousness is at the center of its own universe. The waking, dream and deep sleep states kind of hold their own positions, but when shaktipat occurs we experience what we refer to here in The Hard Light Center, as the *fourth* state – that state of consciousness that becomes aware of the ocean of consciousness.

We have direct experiences of the waking state and the animated perception through the sensory fields. But then we also begin to spontaneously experience consciousness at the level of the subtle body. This is the experience of electricity itself, the electricity of the prana and the life force. We begin to experience consciousness at the level of the causal body which is outside of time, has no space, and is the foundation of mental formation and what we call mind.

There are countless states of mind. We begin to have conscious awareness of this. This begins to kind of tip and rock the boat. Not only that, we will also begin to have experiences of the fourth state, which is the direct apprehension of the infinite ocean of consciousness and the billion suns of force of the kundalini that is the manifestation of the creation.

Sometimes we hear about people that get kundalini awakening without shaktipat. In other words, the kundalini starts to move through them with irresistible force, but the Guru shakti, the transforming energy that causes that circuit to become a useful and healthy energy, is not present, so they'll start to careen through the wild perceptive field of what reality truly is.

The Guru shakti is irresistible and unfailing. Once shaktipat occurs, it cannot fail to bring about awakening. What I've also seen is that there is a steady burning tempo, not too fast, not too slow – just steady – that is one of the main effects of the Guru shakti on the individual.

If a person received shaktipat and then, in the very next second, the entire curtain of maya was ripped away, the person would probably just go insane. They wouldn't be able to deal with it or find balance fast enough. It's better that it's approached by degrees. Just based on my experience and what I've read, when the curtain is finally ripped away everybody thinks it's too soon: "No, I'm not ready".

But then you have to accept it's a destiny. It's what God decided to happen. It's God who says "This is your destiny, this is where it happens for you".

Even though shaktipat occurs, there are still the great forces that we meditate on and contemplate endlessly and rarely understand. These are the forces of destiny and the forces of fate, which are still subtly present inside the unfoldment of every individual. The distinction between the two is very simple. Destiny is irresistible and cannot be altered in any way. It is the manifestation of God as you. But as they say, fate is fickle – it can go this way or it can go that way. Destiny – the destination – is constant. It remains the same and has never changed. But how you choose to travel the path is completely and totally 100% your choice. This is the principle of free will. It is protected at all costs.

Once shaktipat is given, there is a subtle touch of spirit. If you stay in the pocket of the Guru shakti, you'll be moved swiftly to the goal and you will be supported on all sides while it happens, and then inevitably you'll be thrown in the deep end and told to swim. And when you consider that, it's just like everything – the more training the better and the more preparation the better because you want the best results.

Spiritual practices are connected to the way we are put together and the understanding of how the breath interacts with the prana, how the prana interacts with the mind, how the prana matrix interacts with the body and how the body and mind prana-matrix is the support for the apprehension of the fourth state.

The fourth state is not something outside you. When you experience the fourth state you don't see it out there. It comes from inside you, and it comes through you as you.

The Siddha Lineage

There is an entire class of beings that exist in this universe who are the servants of God and the agency of shaktipat. These beings are called siddhas. They are an entire race of beings, some of which have never been human. They come from every level of life, from the highest paradise to the deepest hell. Somewhere along the line, the Guru shakti came and touched them and they were transformed and awakened. They became realized and awakened to the truth.

Many of the siddhas who have been present on the earth once were human, received shaktipat and were awakened. Then, as an aspect of their spiritual training, they continued to swim in the Guru shakti forever. Why? Because it's unbelievably blissful, extremely exciting and very interesting work. And it is this race of siddhas that is one of the broadest agencies of shaktipat that is active and present in this world at this time. Most of them remain veiled and many of them operate in a kind of hiding-in-plain-sight mode, producing an active agency of shaktipat. Once shaktipat occurs, you come into contact with the company of the siddhas, come under the veil of their protection and begin this incredible transformation of awakening.

All the siddhas and siddha agencies of the Guru are completely different. You'll never see two even remotely alike. The impact of shaktipat is very much like the impact of nutrients, sunshine and water on a flower. That impact does not turn every flower into the same kind of flower. Each flower is unique and comes up according to its own conditions and its own karma. You can have two roses, the exact same kind of rose, and they'll grow side by side and they'll be completely different. It's the same with human beings, the same with sentient beings, the same with siddhas and the same with Gurus. They'll have very unique modes of operation that have everything to do with their own nature and the process of how they awakened.

The siddha lineage is the assembly of awakened beings that remain in existence to serve the Guru shakti, to serve God. We've recently seen a shift, which I call 'shaktipat on sight'. Up until even just a hundred years ago there was a kind of conditioning time that was observed before the transmission of shaktipat was given. You'd first go through a phase of working with spiritual practices, which opened up and prepared your system. Then shaktipat would hit you and there would be a very quick trip between that moment of shaktipat and realization, because your system had been prepared.

We are now in an era known as Kali Yuga, which is an age that is defined by the arising of negative forces dominating the positive energies of the world. Because of the confusion of Kali Yuga, the age of darkness, it's impossible to get enough people to prepare long enough in advance to go through the process of spiritual practices, meditation, and opening the system for shaktipat to occur. So the Siddhas basically turned the coin over and now give shaktipat on sight. In other words, shaktipat is given without any preparation. It's simply if the person desires it, it is given. The number of people that have received shaktipat has gone up astronomically. Every retreat offered by the Hard Light Center of Awakening acts as a shaktipat station.

This has been the mode of transmission for a few decades now and it has its good side and its bad side. The hard side seems to be the process of equilibrium. If a person gets shaktipat before they are prepared for it, this dual polarity between the old world view and the new growing world view produces a kind of dynamic tension inside the person. Sometimes the person will long for the old ways or be confused by the onset of a new formation, and there may be a psychological conflict that the person goes through. That has been counter-balanced by an extra follow-up by the Guru and the Guru shakti. It produces a kind of 360° buffer around the person, giving them a kind of protective barrier when they go through that equilibrium phase.

It is a matter of fate and destiny how the Guru shakti comes to you. Shaktipat is its own force. It is the awakening impulse. In other words, it's all that needs to happen. Once you get shaktipat, you're going to awaken according to your fate and destiny. The course of your awakening is going to change according to you and your choices.

In other words, once you get shaktipat and make a certain series of decisions and do a certain series of actions, you'll receive the fruit of shaktipat and awaken completely and totally at such and such a time and place on the map of time and space. But if you get shaktipat and you make an entirely different set of decisions, then you will receive the fruition of awakening and enlightenment at another time and place. The destiny remains the same, but the fate is subtly shifted, based on the weight and balance of your own choices.

The best option is to get realization in the same lifetime that you get shaktipat. Generate as much broken-field running as you can, moving through the unfoldment of your own karma, the unfoldment of your own destiny, and get to that point where gazzzzhhhhhhuuuuum: you hit the Om point and you absorb. But perhaps it will be in the next life or the lifetime after that. Once shaktipat has occurred, death – the separation of the mind from the body – is not significant because shaktipat is running on the principle of absolute consciousness.

The body is always mutable – it comes and goes. It's also the fundamental basis of your confusion. Based on the fact that you've assembled yourself in the waking state and the sensory fields, you have mistakenly taken the body as a refuge of your consciousness. It's a misunderstanding. The fact is you are infinite consciousness. You always have been. You've just forgotten. And this is what shaktipat reveals.

Meditation is very powerful. My Guru, Swami Muktananda Paramahamsa, was big on meditation. He was a yogi his whole life, got shaktipat, and went through a very intense phase of meditation. That's how his awakening came to him. It was very logical that he was my Guru because I also had a knack for meditation. It was very easy for me to meditate, in fact I liked meditating. I think incredibly interesting things take place that are very powerful and very exciting.

The components of the human form that deal with the explosion of consciousness resulting from shaktipat are the brain and the spinal cord. There are three subtle nerves in the subtle body known as the sushumna, ida and pingala, in addition to the system of chakras. The six chakras are located at the base of the spine, the genital region, the navel, the heart, the throat, the forehead and at the crown of the head, where there is a thousand petaled lotus called sahasrara. The sahasrara is the interface between the ocean of consciousness and the creation. It is an extremely complex spiritual structure that sits at the crown of the head.

When you receive shaktipat, it's important not to be afraid. Nothing that happens is anything new. It's more like you're becoming aware of something that has always been happening. Shaktipat introduces a quality of light into your system that blends all four bodies together into a single reflex of apprehension. And it doesn't happen because I say it does, or I want it to. It happens because it's the entire ocean. Who can resist the entire ocean?

For those of you interested in inner space travel, you can notice how the object of your perception and you the perceiver become a single thing. Notice this and the presence of the fourth state will be a fruit in your hand. All you need to do is direct your attention there.

The apparent world of sensory perception is

an external phenomenon and is a reflection of something unseen. The spiritual world is invisible and the underlying cause of the external world. Over the course of many lifetimes in which you have been involved in the momentum of the world of cyclic existence, you have become attached to the phenomenon of external perception and gained a habit of forgetfulness of the inner dwelling source.

The idea of awakening is pulling back the fog or the barrier that has built up through the phenomenal attachment to the external world. This attachment is called karma, which is a build-up of external impressions of the appearance of the sensory world, and acts like a kind of fog or a barrier of clouds. It's as if you were standing on the earth looking into the sky and there was a layer of thick clouds. If you were going on sheer observation you would never know the existence of the sun. It is only when the fog and clouds are burned away that the sun is clearly seen.

This is very much the nature of shaktipat. The transmission of shaktipat to each individual acts as a force of consciousness that transforms and absorbs all of the karma of the existence of the sensory desire-based identity. As it absorbs that karma, it also burns it away. The day comes where the final quality of karmic-based illusion is burned away and eliminated by the light of the Self. The clear sky is revealed and the light of the 10 billion suns appears.

Divine Will

The Guru transmission is very spontaneous. Once the grace of shaktipat is given, the essence of the transmission is a straight induction of pure consciousness. This is also known as the ocean of consciousness, the great light and the Self. It also flows into the appearance of the shakti-based manifestation of the creation, the apparent universe. This is of the nature of bliss, the expression of the creation. Then the aspect of knowledge, jñana, flows into the mind essence and begins to burn off the karma that has come about from mental-based operations. The confusion due to the attachment to the apparent universe in the sensory planes is quite literally eliminated and absorbed.

There is karma that travels from lifetime to lifetime, which is the built-up product of the false refuge in the apparent world and the apparent universe. You live a lifetime of desire-based activity based on the operation of the senses, which is a frequency of desire unto itself. The essence of spiritual purification is the elimination of all forms of ignorance that actually build up in the body, particularly through the senses, which are instruments that reach out and touch the apparent and objective universe.

It's not a negative thing or a bad thing in itself, but what has happened is that they have become so overwrought and so overwhelming that you are left with a sense of forgetfulness of the true underlying spiritual world that is the source and cause of the apparent world. And it is this build-up of karma that the Guru shakti goes after and quite literally absorbs back into pure consciousness.

It's very much as if you are the drop in the ocean and the ocean is the infinite ocean of consciousness. In shaktipat, the ocean penetrates into the drop and all of the parts of what you once considered to be your separate identity are absorbed back into the nature of the ocean. It's like taking a glass of water and pouring it in the ocean. One very swiftly loses track of where the glass of water is and where the ocean is because they commingle so quickly.

The Guru shakti flows into the mind and into

what is known as iccha, the will. The will is the cause of the creation. The will of God is actually transmitted directly into you. During the time that you are in the thrall of the individual identity – you are the drop dreaming you are separate from the ocean – you have what is called personal will.

This goes back to the idea of destiny. It is your absolute destiny to awaken to your true nature, realize your true Self and merge with the great ocean. But it is your fate to find your own pathway there. In this way it's as if we were speaking in parables. Even though destiny is absolute, God has set up the wild card of fate, wherein you are allowed to dream the dream of individual consciousness.

However, it's a matter of destiny when the Guru finds you. Sometimes you'll be looking for the Guru and you won't be able to find him, but he'll find you. Sometimes the Guru will find you before you ever thought anything about a Guru. This happens a lot. Oftentimes a person's soul will be seeking a Guru but their connection to their soul is limited. Their connection to their individual identity so overrides the quality of the soul that the soul will cry out to the Guru to come and the person will be unconscious of that.

Being a spiritual teacher and talking to a person, oftentimes their soul will be speaking and their ego-based identity will be speaking and it's as if you're talking to two people at once, each with different purposes. I'll give you a hint: the teacher always listens to the soul.

This idea of the individual will and the universal or divine will is an especially tender point of relationship because the idea is to get the individual to seek their own true Self. Even though they are a drop in the vast ocean, they are allowed to have the dream of individuality as long as they want it – up to the moment of destiny when they awaken. And so this idea of individual will is worked out between God and the individual, through the Guru.

You hear this phrase "God, Guru and Self". God and Self are on the two ends and Guru is in the center. They are not three things. God is the Guru. God is the Self. It's very much true that the Guru principle is the bridge, but it's not a bridge from inside of one's self to outside of one's self. It's a bridge from inside of one's self to the essence of one's self. This is the nature of the Guru shakti and this is the nature of shaktipat. It eliminates this barrier of desire-based activity and karma, which has built up in such a way that it produced a forgetfulness of the connection to the divine Self.

As the Guru shakti eliminates this barrier, it reveals what is apparent. Of course, once it is revealed, there is never any doubt. As soon as you experience the true Self, you feel that you have come into direct contact with God and you know that God arises within you as you. There is no doubt. It is not a philosophy that can be questioned.

There are countless beings that will happily choose their dream of individuality. This will be allowed to unfold for quite literally thousands upon thousands of incarnations, until one begins to suspect that something unusual is afoot. Then one will begin to generate a yearning for truth and this yearning will be a seed. You will become a person who would rather know what is true no matter what it costs. You'd rather know the truth than to live in a lie or an illusion.

This clarification of mind begins to affect the individual will very intensely. Even just a fundamental intellectual analysis of your existence in this world clearly reveals that every individual is connected to every other individual. The fate of the one is the fate of the all. All things are interconnected. We know this to be true from basic physics. We're all made of the same material. We just have to go from one level of perception to another. Just from logic, you can clearly begin to apprehend the inner-connectedness of things. And this leads you to a set of ideas that affects how you feel about yourself.

Almost without exception, what you'll see in a person that's at the turning point is how that person begins acting for the benefit of others, leaving aside their own self-interests at every turn. A child will act almost exclusively for their own self-interest. It's always a very interesting point the first time you see a child do something unselfish. It's very remarkable that they felt it within themselves and they did it. This idea is very powerful and it's actually one of the giveaways that a person is actually beginning to be a true human being. They come to understand the inner-connectedness of everything, and their understanding overrides the instinct of their personal will, so they can act from a higher basis.

When you begin to see a person acting in selfless service, their actions reflect service to the greater benefit of the whole and to the greater benefit of others. They find a sense of fulfillment in that service. They don't feel like they are diminishing themselves or taking a part of themselves and giving it away, because that's impossible. The reverse is actually true. It's a false understanding to think that you only have so much and you have to hold on to it with everything you have, otherwise if you let go of one bit of it, then it's going to go away from you and you will be diminished. Rather, the higher understanding is that by giving your light to the sum of light, light is increased.

As you awaken, you begin to sense the presence of the divine will. It appears as a kind of intuitive knowledge, in which you know what the right thing to do is. You go through your life, you take care of business, you do your job and you take care of your loved ones. Then you might find yourself in a position where the entire situation is a question mark. And the next thing that happens says everything about you.

This idea that I'm talking about also has an even greater over-arching idea that is called right action, and at its highest expression is characterized as skillful means. This is the level of the action of the enlightened, where – without thought, without consideration – a person is so attuned to the truth of a situation that they can act absolutely in the spirit of truth. Their action is so perfectly executed that what was supposed to come out of the situation exactly happens. It's actually a condition beyond the mind. That's why the idea of skillful means is considered an enlightened expression of action because it's beyond even thought.

Acting from the vantage point of right action means that one performs every action without the slightest error, without the slightest taint – perfect behavior, perfect action, perfect motivation. This is of course the ideal. It is in action where the rubber meets the road and that's where the acid test is. It's not theory. You can't have enlightenment as an ideal and behave in a selfish manner. It produces a collision and something's got to give, and that produces a crisis of will. This is often times why once a person gets shaktipat and enters spiritual life, they feel like there is a pressure pushing them, which seems to come from everywhere and nowhere. Its source is the ocean of consciousness itself, which arises as perfect consciousness and skillful means. What is present inside that envelope is your own enlightenment.

You'll find that you'll reflect into enlightenment, multiple times through your day. You'll feel the presence of God inside you and sense what is supposed to be happening. And you'll see where you've had the courage to commit to aligning your individual will with the divine will, and use every aspect of your personal power to execute the moments of your life. You'll find that the layers of life are not just out here on the external – which is actually the slowest moving part of the apparent physical universe – but that we also have the electrical charge of the life force itself, the prana, and we have the absolute omnipresence of mind and mental formation. You have to become conscious of all of those parts of yourself to make this happen.

Before shaktipat you would never have enough energy in your system to be aware on all of those levels. It would be almost impossible. But the Guru shakti is like a divine hand. And yes, it poses the question, *"Who am I?"*

This is shaktipat and its fruit. It comes with the entire force of the ocean of consciousness. It is irresistible and cannot fail. As I said before, it is not the operation of the personal power of one individual but the entire siddha lineage expressing itself in full force. It is something extraordinary.

CHAPTER THREE – HOW

The Guru is an eternal spirit that appears as the power, grace and mercy of God. It is the very empowerment of enlightenment itself, appearing in countless forms, with great force and concentration through time and space, and arising in the heart of every sentient being. The Guru is the bringer of rain and the bringer of grace, setting into motion the wave of awakening that is spontaneously born inside everyone and everything. This wave of awakening is known as shaktipat, which means 'the descent of grace'. It is in shaktipat that the initiation of remembrance and realization is set in motion.

It is often said that the Guru lights the flame within you in the same way a flame of a candle is lit. It must be understood very clearly that the Guru dwells within you as you and that it is this touch of shaktipat that is the essence of the Guru's speech. The Guru brings the light of awakening, and it is this activity of awakening that is called dharma. It is this dharma that is the path or way. Having received the touch and the light of the Guru, we learn to live and exist in that light so that it is always strong and grows ever brighter.

In the Vedas, the original Guru was named Vyasa. He is an arch-personality that is defined as that spiritual force that brings the power of the spiritual laws to the universe. It is of the nature of Vyasa to translate the divine law of God to the world, translating it into that language of countless sentient beings and world systems.

So in this way the personality of Vyasa is the arch-persona of the Vajra Guru, the primordial presence of illumination and awakening, appearing not only in this world system but in countless world systems and supraconscious personas.

It is said that there have been twenty-eight incarnations of Vyasa in the time since the beginning of our world system here, some 400,000 years. These incarnations have made it possible for the incomprehensible speech of God to be heard.

This is the nature of the Guru, which is an

incarnation of mercy and grace. Our world system is named Endurance. It is endurable because of the presence of dharma and the presence of the Guru. I oftentimes say that the presence of dharma is what saves the world.

Everybody always says to me, "Well, it sure doesn't seem like there's much dharma in the world", and they've got a point, but what dharma there is makes the difference. If the dharma was extracted from the world system, then it would not be endurable. It would be a matter of inconceivable suffering with no escape.

It's interesting that it is only in our awakening and in the expansion of our illumination that we become truly aware of the kindness that we have been in receipt of, bestowed through the power and presence of the Guru and the Guru shakti, and the vast assembly of awakened souls that have remained in existence and have become the servants of those seeking awakening. We refer to those beings as the Guru lineage, the association of awakened souls. They operate and act in concert, generating and establishing the awakening force, maintaining its presence here on this planet, so that it can be reached.

Because of their service, there's always a room or door someplace that you can walk through, meet the Guru and receive shaktipat, the awakening touch.

The Guru Margena - The Path of the Guru

I want to discuss in detail the internal divine structure and architecture by which a human being becomes awakened. The essential points are spoken of very clearly in the ancient text of the Guru Gita, which gives a description of what is called the Guru's path, the Guru margena. It is the Guru's path by which shaktipat is generated. The Guru margena, the Guru's path is spontaneously present in the human form.

This refers to a very profound idea that birth into a human form is extremely rare and difficult to attain. It's also considered to be a final form. They say that even the gods envy a human being because this Divine architecture is present only in the human form.

It is said that to see a human being in all the expanse of the three worlds is as rare as seeing a star at midday. Having done some traveling, I can tell you that is true. The human race is an extreme minority.

Non-human sentient beings fill the three worlds like a blizzard. Once you travel off of this world system, it can be a long time before you will see a human form. There are other planets that give rise to the human form – some 1,800 in this little quadrant of the universe. There are universes of universes. If you've heard that, it is true. After awhile, it starts to lose its meaning. Just like in the Wizard of Oz, there's no place like home. It's because home gives us a context – home being the endurable and the beloved.

The Guru Gita is a very profound mantra. Essentially it is the outline of how to do sadhana – how to become awakened. Simply read it as an instruction manual. It couldn't be clearer. It presents the understanding of the Guru margena, the path of the Guru, and tells the seeker to follow the path the Guru reveals. The words are poetic sounding, but what is being referenced is a piece of divine architecture – a fiber of light that is present inside the human form since birth.

The path of the Guru margena is a subtle fiber of light that flows from the crown of the head to the secret space of the heart that is the seat of the Guru. This is the pathway of shaktipat. It is the Guru margena that is illumined and activated in the process of shaktipat. A surge of light is sent down this fiber of light. It is so profound and so subtle that it sends a reset signal to the entire human form.

Unfortunately, due to the way the design has occurred, the senses have become too empirical. They irresistibly load their impressions into the brain with such force that they override every other subtle impression. The awareness of the subtle body, the awareness of the pathway of the three rivers of ida, pingala and sushumna, the awareness of the divine vibration of the Guru margena and its architecture of the six wheels of the chakras, are all veiled behind the activity of the senses.

Many people have read about the astral body or subtle body in books. Maybe they've felt a tingle or a throb of energy there, but because of the intensity of this energy that is hard-wired to the senses, too much of the brain is dominated by that force so that it simply overrides everything.

An incredible storehouse of impressions of the five organs of action is built up and assembled along the avenues of the three principle rivers or fibers of the subtle body: pingala on the right side, ida on the left side, and sushumna in the center.

The impressions are also assembled in the chakras of the base or root chakra, the second chakra, the navel chakra, the heart chakra and the throat chakra. The sixth chakra is unique because it draws energy from the sahasrara and the energy of consciousness at the crown of the head and above. It also draws energy from below, from the five elemental energies of light that are connected to the five lower chakras. The sixth chakra is like a dashboard in the middle, between universal consciousness and the appearance of the world.

When the mind is ruled by the senses, everything from the sixth chakra down overwhelms the mind. The senses operate inside the field of time and are connected to the physical body. But even the subtle body, as powerful as it is, is pressed into the unconscious by the over-arching power of the sensory fields that electrically dominates the brain.

From the moment that you emerge from your mother's womb and your senses switch on, impression after impression pours into the brain from every direction and is stored magnetically along the subtle nerves of the ida, pingala and sushumna.

The sensory field is ruled by memory, which gives rise to the intellect. The intellect is the ordering power between the senses and the memory base, which uses pattern recognition for discrimination. For example, it is the intellect that says, "OK, there's something green. What's green? Grass is green, trees are green, green chilies are green, that looks like... oh, it's a tree." And you go through an association chain in your intellect, and you find out that it's not this, not this, not this, but yes, it's a tree. And you think, "I've seen a hundred billion trees. What kind of tree is that? Oh yes, that's an oak tree." That's the empirical relationship between the intellect and memory. What has been seen before is matched up with what is being seen now. Because you've seen a lot of trees, it's easy to define a tree, based on other trees you've seen.

But how often have you seen the pure white light of eternity? How often have you seen the nature of the void? How often have you seen the fiber of the kundalini surging through your subtle body? How often have you seen pure consciousness, which is beyond all description? How often have you seen the subtle worlds that are aligned with the subtle body?

You may have seen them, but you haven't seen them as many times as you've seen a tree. So your memory base is thin when it comes to subtle physical phenomena. It's not that it's not there. It's not even that you're not seeing, but you're only seeing and recording it subconsciously and unconsciously.

Why? Because the brain is dominated by the empirical order of sensory phenomena, memory and mind. Think about it: how many impressions have happened just in this lifetime since you were born? And you've recorded everything perfectly.

This whole idea of the Guru lineage is that the Gurus are the travelers. They've been through all that and discovered that there's more beyond. At one point or another in their flow of incarnations, they've disappeared from this world and travelled in the countless worlds of consciousness and light. They've done this so many times that they have incredible, overwhelming amounts of impressions of the inner worlds – not only the physical envelope of the world of appearances, but also the subtle physical world of the realms of light.

When you read the text of the Guru Gita, which is a simple description of the awakening process of the human being, it says 'seek a Guru'. Furthermore, seek a Guru that has had a Guru. This chain of interconnection speaks to a depth of experience of those beings that have accomplished awakening by virtue of the Guru's grace and by virtue of the Guru margena.

This fiber of light that I refer to as the Guru margena is simply that – it's a fiber of light going through the center core of your being, from the crown of your head to the secret space in your heart.

It's not difficult to illuminate the three relative bodies of the physical, subtle, causal body of mental formation and finally the fourth body of pure consciousness. As in anything else, you simply have to learn how to pay attention. It is always a matter of attention – and also concentration and meditation. In the same way that operating within the waking state of the physical body requires the capability of concentration, operating within the subtle body requires meditation. It also requires the discipline of discrimination, which is the ability to discern the difference between subtle phenomena. They're not just spiritual skills. Those beings that have strong powers of discrimination, discernment, concentration and meditation in the physical world are capable of accomplishing a great deal.

The fact is, one of the swiftest ways to awaken a human being is to awaken and illuminate the subtle body. The light inside the architecture of the subtle body will force perception to change. You can talk to people until you're blue in the face, and all they'll do is argue with you. Believe me, I've tried. But if you simply turn the light on, that forces a change. This is why shaktipat is referred to as the fast path. It's not a path of belief, but one of direct experience.

This is a way that works. It has worked countless

times, and it will work with you. You don't have to re-invent the wheel. You don't have to invent a way. The path is given, it already lies within you.

What I'm describing is this divine architecture of the Guru margena, which the Guru Gita places up front as the means of awakening. Shaktipat involves the firing of the kundalini into the Guru margena, and awakening it – switching it on. In an unawakened condition, it sits in a state of potential. It is dormant. The sushumna goes from the base of the spine to the crown of the head. It's shaped like a conch, like a spiraling structure. It's not just one current – it's like complex streams within streams.

The Guru margena is concentrated in a very focused way from the crown of the head to the secret space of the heart, which is known as the akasha of the heart, the seat of the Guru. In spiritual life, this entire structure is considered to be one thing. In other words, don't think of it as the heart chakra, and then the throat chakra and then the sixth chakra and then the seventh chakra of sahasrara. The structure of the Guru margena is all one unit. The seat of the heart to the apex at the crown of the head is a single component.

The Rupa ~ The Blue Pearl

In the human form the infinite consciousness takes the form of an infinitesimal blue dot, like a blue atom, which in the Guru Gita is known as the rupa. It has also been called the Blue Pearl by the siddhas and sages.

Inside this blue atom is the infinite seed of creation, the universes of universes. You can catch a glimpse of it on your approach as you begin to build up spiritual force through the conduct of the Guru margena and the power of shaktipat. In the Guru Gita, it is said that even a glimpse of the Blue Pearl grants liberation, because once you've seen it, you've seen the totality of the truth distilled down into a single atom.

You'll especially feel it moving in the secret space of the heart that is the seat of the Guru, up through the Guru margena and into the brain. You'll be seeing it in every kind of situation – inside beings and inside yourself. It can't be summoned. You have to just become perfectly still and aligned with the truth.

The Guru Gita also speaks about rupatitam, which means beyond rupa. This is where the rupa, the Blue Pearl, itself is shattered and form arises as empty. This is a way to describe the moment of liberation, when one is no longer bound by any form of karmic compulsion. The mind has been stopped, the intellect dissolved, the ego annihilated, and one's consciousness is assembled out of consciousness itself. It always was assembled out of consciousness itself, but it is the power of the modifications of the mind, which were of course only illusory, that has made us think otherwise.

The rupa, the Blue Pearl, is that amalgam of consciousness that arises both as form and the formlessness of the rupatitam, emptiness. It can only be reached by the Guru's grace. These aspects of existence are of the deepest, most profound and most subtle nature of being. It is in the awakening of the Guru's path where all of the components of the divine architecture are kindled and illuminated, activated and empowered, and then finally integrated. And in that integration, an extraordinary power emerges.

The spiritual world is opposite the physical world. In the physical world, the greater the mass, the greater the energy. In the spiritual world, the more subtle the mass, the greater the energy. The rupa, the Blue Pearl, is one of the subtlest forms of phenomena, and it carries the energy of the entire creation. The rupatitam is the capacity to perceive the emptiness out of which it emerges. It is in these states that samadhi, the direct intuitive apprehension of the Truth, becomes possible and begins to occur.

The Structure of Sahasrara

You can do your entire sadhana from the information that's in the 108 verses of the Guru Gita. Everything is touched upon. These 108 verses describe the entire essence of the Guru margena, the path of the Guru and the path of awakening from shaktipat to realization. They give the description of the unfoldment of awareness, carrying a powerful transmission about the Guru margena, the essence of shaktipat.

It's important to read the Guru Gita with an analytical eye because there's an extraordinary amount of information there. As the vibration of grace along the path of the Guru becomes more profound, the force of the kundalini will grow, the presence of the SoHam will grow, the direct apprehension of the divine rupa will grow, and an awareness of consciousness beyond form – what we call emptiness – will grow. That is as much of enlightenment that can even be described or talked about – quality beyond all qualities. I'm focusing here on the divine architecture of the human form, which is what makes shaktipat possible and available – how it functions and why.

The description of the four bodies and the unfoldment of all of the different conditions of life, are all addressed in a few condensed sutras in the Gita, each one packing a very powerful wallop of meaning. I want to focus your attention on the essential lines in the text, which are the absolute essence of the Guru Gita, the essence of shaktipat, and the essence of the Guru's path.

> *Residing in the center of the thousand petals is a divine triangle formed by the Sanskrit alphabet, with the letters A, Ka and Tha at each point. One should meditate on the Guru's two lotus feet, which are Ham and Sa, in the center of this sacred triangle. v.47*

This is the description of the three-pointed triangle sitting flat at the crown of the head. This triangle is underneath the skull, in the very fabric at the top of the brain. It's inside the contour of the body. This is the Guru's form, with the three corners A, Ka, and Tha. It sits just under the brahmarandhra, the soft spot at the crown of the head. It's inside the contours of the physical body, and laid into the cerebrum, the very top patch of the gray matter that has to do with the highest brain functions.

The essence inside of the triangle is void, with the mantra Hamsa arising at the Guru's feet, which are the syllables So and Ham. The syllable So is the vibration of the appearance of the creation, the manifestation of the universe, and has its seat at the base of the spine. The syllable Ham is the vibration of the ocean of consciousness, beyond form, beyond quality, and of the nature of pure consciousness – empty in nature, but infinitely conscious.

You'll oftentimes see reference to the SoHam, which is the primordial vibration that spontaneously arises within the human form, referred to as Ham and Sa or Hamsa. It can sometimes be confusing because they are not two different mantras. The Ham and Sa are the SoHam. Those two syllables dominate when the mind is externalized, when the mind believes in the senses, and the mind believes in the intellect and the I-consciousness. Then that vibration sounds like Hamsa. But when the consciousness internalizes into the innermost stream of the sushumna and the Guru margena, it becomes SoHam.

This triangle structure at the very top of the brain is the interface between the crown of the sushumna as it goes into sahasrara. The divine architecture that is solely dedicated to the highest operations of consciousness is unbelievably profound and elaborate. It's where the action is.

We're kind of like an antenna. When you see an antenna, you'll see a long column of metal starting at the base and going up to a fine point. You'll tend to see apparatus at the top that are designed to pick up different kinds of waves. Sometimes you'll see them at the middle, or just down from the top or at the very top. They're like dedicated circuits designed to pick up different bandwidths. The sahasrara is like an antenna that picks up the bandwidth of the purest form of consciousness.

When you have an intent, what determines the power of that intention is how much electricity is going through it. Say you have a radio antenna – if you're just plugging it into a wall socket that doesn't have very much electricity, it doesn't have much of a magnetic field and your radio doesn't pick up much. What it does pick up is pretty filled with static. But if you hook it up to a normal generator, or if you get a bigger generator, or if you hook it up to the Hoover Dam... all of a sudden you've got an antenna that can pick up everything that's going on in the planet earth – even stuff that's going on in outer space. Why? Because there's so much energy moving through it that it can magnetize any level of activity. You can pick up energy from a star twenty light years away. The concept is the same with the sushumna and the sahasrara.

Before you get shaktipat, you're just kind of plugged into the wall socket and you're doing the best you can. You'll never have enough energy to draw down that infinite energy and make use of that incredible apparatus. Once the Guru margena is switched on, that sends the thunderbolt down through the fiber of the sushumna, ida and pingala, and clears all of those subtle seeds of obscuration that are sucking energy away from the sushumna, and in all the subtle seats of power within the subtle body. As these are cleared out, your system becomes increasingly more efficient, so you are capable of generating an enormous amount of energy from the base of your spine to the crown of your head. As that happens, all these wheels or chakras begin to kick on, and that clears all of the stuff even more. Eventually there are no more stored karmic seeds of obscuration. Then it's just pure consciousness that conducts itself along the antenna of the sushumna, ida and pingala.

As all of this occurs, an enormous amount of energy begins to build up in the sushumna, and the entire conduit of the antenna from the base of the spine to the crown of the head becomes full of energy and vibrates. This gives you the capacity to jump the gap at the crown of the head, merge into the sahasrara, bring that incredible apparatus online and begin to draw the energy down. It is the sahasrara that is the complete dedicated circuitry for conducting the vibration of the highest level of consciousness.

This gives you the capacity to perceive the most subtle phenomenon, just like a radar telescope can subtly discern the atomic formula of a star a hundred light years away. It has enough energy and it's a subtle enough ray of consciousness to discern that truth. This is what becomes available when the full force of the sushumna, ida, pingala, the Guru margena, the A, Ka and Tha triangle, and the seat of the Guru's feet at the crown of the head are fully empowered, drawing the sahasrara online. Then you can perceive the subtlest phenomena in the creation, which is the rupa, the Blue Pearl, subtler than the subtlest. This gives you the capacity to discern the most subtle aspect of phenomena.

As you gain the energy and capacity for internalization, the Hamsa becomes the SoHam. That means all of that force of creation is internalized and moving in the central nerve of the sushumna, ida and pingala. By the sheer conduct of its movement along these fibers, it is purifying, dissolving and eliminating all these different seeds of obscuration.

As it merges into the central channel, you gain the capacity to switch the senses off and stop the mind. With the stopping of the mind you gain an enormous amount of energy that can then reach to the crown of the head where there's a vast magnetic field.

You have to understand, the apparatus of the sahasrara at the crown of the head is just unbelievable. It is unbelievable! It has infinite capacity to conduct the energy of the consciousness of God. As the obscurations are purified, you can then hook up to this apparatus that is the sahasrara, and draw that energy down. When that happens, it's like going the speed of light. You have the capacity to break the bonds of the binding power of the wheel of samsara, break the boundaries of earth, and move into infinite manifestation. You suddenly arise as everything.

All of the heavy-gun activity of emptiness originates in the sahasrar. It's not so much that God is above and we are below, but the structure of our dedicated circuitry in the human form produces that impression. It seems to descend from on high, come down and penetrate us – and indeed it does, because this is the antenna array for emptiness. It is in this space that the seat of the triangle, which is the form of the Guru within the human form, is located.

The three-pointed form of the triangle corresponds to the three sacred places that assemble the three worlds: emptiness or pure light, energy or speech and form or action. These seats are at the forehead, at the throat and at the heart.

There is another verse that says that the essence of the Guru resides in the third body, the causal body, which is present at the heart. This is a supremely sacred and secret place.

> *In the heart is a cave the size of a thumb, which is the seat of the causal body. Listen, and I shall speak to you of the meditation on this form of consciousness. v. 49*

There are one hundred fibers that originate in the region of the heart, fan up through the array of the brain, and divide across the five formations of elemental energy: earth, water fire, air and ether. In other words, they comprise all relative formation.

This is where karma is assembled and all the subtleties of experience take place in these arrays. They are the circuitry for everything that you experience as yourself.

These hundred fibers are aspects of ida, pingala and sushumna. The A, Ka, and Tha triangle is an anchor for these three rivers. There's a division of activity of these hundred fibers. What we experience as existence is the interaction of consciousness as it moves in these hundred fibers. This is hard-wired circuitry – consciousness moving through form.

Of these hundred fibers, there is one fiber that goes from this secret seat of the Guru within the causal body, that is a hundred thousandth of a hundred thousandth of a hundred thousandth of the size of a human hair. As you draw energy to this place, you'll feel the energy moving up and flooding the brain. It will purify all the fibers.

As the Guru Gita says, in the region of the heart there is a structure shaped like a cave, the size of the thumb. Inside this structure is the secret essence of the Guru. It also is the anchor of the one single thread that goes through all hundred fibers, up through the very center, bypassing all the hundred fibers to get to the triangle of the three points, A, Ka and Tha, and up through the

Hamsa. So it goes from the hundred fibers, to the three, to the one. The one fiber that threads through is the Guru margena. It goes down from inside the Guru's seat of the three-sided form, the Ham and Sa, and from there, the one fiber threads the needle.

The Sufis call it the reed of God because it is one fiber out of millions of fibers. There are 72,000 fibers branching off the three fibers and the six chakras. Of that 72,000, a hundred are important: there are fifty for the male side, pingala, Ham; and fifty for the female side, ida, Sa. As it hits the heart, it spreads out like a fan. All the data of your experience is stored there. It is where the experience of the wheel of cyclic existence occurs. But out of that hundred, there is one that originates from the secret seat of the Guru at the very center of the body.

That's why they say that the Guru is the fast path. All these forms have to be illumined, activated and purified. But working from the outside in, you would need about 20,000 lifetimes of non-stop effort. The Guru's grace saves you the trouble and says "Here's the one fiber that counts, follow this awakened path, hold to the Guru's line, hold to the Guru's path." This is the Guru's path that is talked about in the Guru Gita as the Guru margena. It's not out here. It's not like, "Go stand on your head, hold one arm in the air, stand on one foot" – that's not the Guru's path.

When shaktipat is given, the Guru comes along and strums that fiber like a guitar string, so you feel it. Pay attention here, this is the path. I know this is true because this is what I did. Allow it to be pointed out to you and pay attention to that. You will rise up that path, you will fly through the crown of the head like a bird flying through a skylight, and you'll plug into the true universe. All of this that you see out here is a reflection of the true universe that you discover within.

This is where form is transformed into formless emptiness. From the crown of the head down is vibration and formation. From the crown of the head up, it's void, emptiness. The vibration of the Guru margena, this thread, this reed, is empty. That's why, without the Guru's grace, you'll never find it – it's empty. Your mind can't recognize what it has never seen. What does something that doesn't exist in any way, shape or form, look, sound or feel like?

That's why they say that the Guru is the means. You have to have the Guru. Why? Because he's the one that strums that string. That's what he does for you – it's a kindness. It suddenly moves, and you say, "Woah! What was that?" It doesn't feel like anything else – it's a whole other thing and you can recognize that instantaneously. Why can you suddenly see it? Because it's moving and has been made active.

That's shaktipat. It can be activated with a thought, a look or a touch from the Guru. Shaktipat is the ignition and activation of the Guru margena. Once it's activated, it generates a wave of consciousness that sets everything into reverse – reset, dialed back to zero. It is a movement of consciousness that has nothing to do with belief.

But it's also important for the recipient to be paying attention appropriately. The results are better. Otherwise it's like I keep pitching balls and you keep missing them. If you know it's coming, you can catch it. The vibration occurs whether it is recognized by the conscious mind or not.

Each time shaktipat is given that vibration occurs. It vibrates through the entire structure of the human form, through all three relative bodies – physical, subtle, causal, as well as the fourth body, the supracausal.

The ability to align with this true Guru margena is your salvation. After you learn to pay attention to it, boom – you're off and running.

The subtlety of this one fiber is beyond description. Without the Guru's grace, you would never find it. It would be as impossible as saying: "There is a needle, and I've hidden it in North America. You have to find that needle and here's a clue."

The Guru margena encompasses the entire structure, from the three-sided triangle at the crown of the head to the essential bindu at the heart. It is the one single fiber that traverses this distance, penetrates at the crown of the head and captures the bindu, which is the first seat of realization at the crown of the head. It's about a quarter of an inch above the skull.

When you chant the Guru Gita, that thread becomes magnetized. You just have to look for it. You'll find that there's this magnetic throb pulsing down this infinitely refined fiber. That's the Guru. It is the one fiber that penetrates through all four bodies and moves into the space of emptiness.

Where Does This Lead?

Spiritual life is very profound. From the standpoint of the ego, it's a little bit scary because the ego is annihilated. People often ask, "Where does this lead? Why do we do this?" That question always comes from one place. Oftentimes, a person will begin to move along this path with great force and find themselves plummeting to the source of where the Guru margena begins and ends. They know for a certainty that is death. Everything that they have been or have thought is eventually going to end and they don't know what the next thing is. Almost invariably, the first time someone experiences deep meditation, they go, "Whoa!"

It's a natural reaction. We don't talk about it much: "Oh yeah, profound meditation – it's just like death." Meditation would be even less popular than it already is.

But you can look at it in another way. You're going to die – it's a certainty. Your mind will be separated from your body. It's such a panicky moment and then the next thing you know, you're in the Bardo, where you're experiencing the judgment, then the wrathful deities come after you, and then the next thing you know you're just praying for another womb to be born in. Then you forget everything and boom – you're born again. You have no idea how many times that has happened to you.

A good way to understand deep and profound meditation, also known as samadhi, is to see it as practice for the moment of death. You go into a deep meditation, and it's like going into a deep cycle of annihilation, but then you appear back in your body. You don't have to be reborn, it's just you again. You're in your room, you're crashed out against the wall in your kitchen, three days have gone by. But you can get up, get in your car, and go get a burrito – it's great.

You start to become conscious of that space between birth and death. It's there, you've been through it a thousand times, ten thousand times, even more. Why can't you remember? It's the concept of memory.

We have subtle prejudices because we believe we're an "I", we believe in the existence of the body. We think that if the body is alive, then we're alive, and if the body is gone, then we're not alive. It's a false refuge, it's a misunderstanding, it's a confusion. What we know is the Bardo takes place in the hundred fibers that arise out of the Guru margena. What transfers from life to life is the mind. It's the mind that stays constant.

There's a very humorous moment when you start to go into samadhi and there's some of the trauma of annihilation of the ego – it's just the ego that's annihilated. It's not even keyed to the body because in profound meditation, you can go into samadhi, annihilate the ego, but keep hold of the body. The yogis have proven that and that's what we're learning how to do now. It's just a subtle shift of attention and understanding.

One of the perks of going into deep samadhi is gaining the memory of your past lives. Where are they stored? They're stored in your subtle body. You don't have access to them now because you don't really believe you have a subtle body. Or perhaps you now believe it, but it also takes training to learn how to access the subtle body.

One of the benefits of profound meditation is that you begin to see. Not only do you get to see who and what you've been, but you also get to see who and what you will be because the past and the future are both equally illusory. One of the funny things is that you don't change that much. Your character changes. But you would recognize yourself if you met yourself on the street in another life. Nine out of ten lifetimes, you look pretty much how you look right now. Why? Because your mind's pretty much the same. It's the content of your mind that affects the look and appearance of your body. It's the mind that has to change to generate true and profound spiritual change of character.

One of the basic propositions of spiritual life is that you have to learn how to act out of generosity and you have to cease acting out of greed. You have to learn how to act out of love, you have to cease operating and acting out of anger or hatred. You have to put a stop to jealousy and envy. You have to put a stop to fear. It has to happen.

In spiritual training, one of the first things that must be taught is about dispassion and detachment. These are the tools that you use to shape or get a grip on your behavior. As a spiritual teacher, it's not that difficult to illuminate a person and activate the Guru margena. What's difficult is to get someone to improve their character, because that conversation is always between you and God, between you and the Guru. It's a constant conversation, it's a constant battle. All these seeds of obscuration have their origin in ignorance, desire, anger, jealousy and fear. As long as these tendencies exist, they subtly and subconsciously assert a conditioning factor on your mind. This is why we begin to do sadhana, why we learn to conduct the fire of the kundalini shakti through the ida, pingala and sushumna, why we intensify the awareness of the SoHam, and why we internalize our consciousness – so that we draw it off the externalized constant ensnarement of the six sensory fields, which are the most profound trap of maya and the wheel of cyclic existence.

First we use the rule of opposites, which is like

using a thorn to remove another thorn. When we experience anger, we give rise to kindness. When we experience jealousy or greed, we give rise to generosity. We get in the habit of producing the opposite, to at least cancel out the effect of the negativity.

There's a point where the effect of the kundalini and the discipline of leading a spiritual life eliminate these incredible storehouses of obstacles. The movement of the SoHam in the sushumna, ida and pingala purifies all of these impurities and obstacles. Every time the throb of the kundalini moves in the central nerve, it eliminates oceans of these impressions.

This is why continuousness is always advised. Pay continuous attention to the mantra of the SoHam. Pay continuous attention to the presence and power of the Guru. Pay continuous attention to the conduct of the kundalini shakti moving through the central nerve. These are the mechanics of purification.

This is the value of continuousness: maintaining a constant bead on what is important. It's not difficult to do. You don't have to eschew the world and sit in a cave. In fact, it's better if you don't, because that sets up a wave of extremes. The middle way is best. Go about your life, meet your responsibilities, but in every aspect of your life and in every breath of your life, see the SoHam and remember the Guru.

When you see into the seed of all the forms that come at you in your life, know that it's just the unfoldment of your own karma. Much of it has to do with maya, but if you see through to the heart of it, then it will vanish like a mirage. Indeed it will. If you're captured by its fear, it's got you. It's like a demon or a vampire – they can't come into your house unless you first invite them in. It's a beautiful metaphor. Refuse to be controlled by maya and it becomes powerless. You have to learn to become fearless. Place your faith in the truth.

I want you to understand that the Guru margena and the path of awakening are something you're born with. It is the Guru that throws the switch, but the path is already there.

Just in this time that we've been placing our attention on the Guru margena, you can feel this incredible presence at the crown of the head, flowing through the innermost stream of being. In Guru yoga, this is what we're describing. As your Guru yoga becomes more profound, your being will become flooded with bliss. This is a never-ending experience, as much as you can bear and more.

Perceiving the True World & Emptiness

Shaktipat then is the awakening of the human being and the fast path, the direct manipulation of the subtle body, which forces the mind and the body to open up to its own true reality and to its own true indwelling inner Self. Shaktipat is the presentation of an alternative to the wheel. In fact, that's what we do when we get together. We argue about reality and I present an alternative. This is how Guru yoga works.

By reciting the syllables of the Guru Gita, that path, that fiber, the true Guru margena, vibrates. It's so powerful and so clear that everything else is silenced. You learn to align your attention with this subtle form of speech. It will transform you. Once the Guru margena becomes active, that vibration will heal, override and absorb all illusory appearance. Again, this is why the Guru is the fast path. This is the way it occurs.

Through repetition and habit, the conscious mind is trapped in the phantasmagoria of the appearance of the physical universe. It is enthralled with the six spheres of sensory perception, which give rise to confusion. In reality, that which is not true can be seen; that which is true, is invisible. An entire reversal of the basis of experience must be undergone. That's sadhana.

Until your eyes open, the true world is invisible. As you awaken, it will reverse. You'll live in the true world and you'll see this as a shadow-play. That cross-over of perception is a little unusual and confusing, but everybody goes through it and you'll be fine. It is of the nature of the world that you are born with a shadow over you. It is the Guru that has the power to remove that shadow. Even when the spiritual masters are born, they are born with a shadow over them, and that shadow has to be removed. That's the role of the Guru.

Even the spiritual masters deal with this. As they're born into the envelope of the physical world, they take on the shadow of that envelope. The Guru must remove it before realization arises. But there's this kind of wild reverberation. As the shadow is removed, the identity goes through the adjustment of the false-identity attachment to the physical world dropping away and the revelation of the true world dawning.

Usually we see that in the incarnation where the human being first experiences shaktipat, there's a kind of overlap. The reality of the past impressions and habits of mental formation, the belief in an "I" and the operation of the sensory fields have a kind of momentum to them. At the same time, the vibration of awakening that is kindled by shaktipat and the path of the Guru, the Guru margena, run concurrently. You'll see a lot of upheaval because these colliding belief systems are taking place inside the being. Don't let that throw you. It's just the power of the change.

You can actually tell the number of times that a spiritual master has returned in the case of a multiple-returning soul. The first time awakening happens, it's an extremely dramatic and disorganized affair. But like anything, after you've gone through that process several times, there's less and less disorganization and disorientation, and the soul then becomes capable of very powerful action.

Baba Muktananda was such a soul. He was an extremely talented being who carried the great light of shaktipat to the West. For that, I can never thank him enough. That's how I met him. How kind that he came to me. His poster showed up right outside my front door. I thought it was an ad for a jazz musician. I actually walked by it for a week before I figured out what it was.

The harmonious attention to the movement of grace in the Guru margena brings about a divine life of cooperation with this divine impulse. To the extent that you give favorable attention to this form of grace, is to amplify its speed and its effect. As Baba Muktananda would say, and as Kalu Rinpoche would say, always remember the Guru with love. If you do nothing else, then his grace will unfold. If you understand these internal mechanics, you can do yourself the favor of accelerating your movement along the path.

Several of the verses in the Guru Gita refer to the advantages of having a guru. With the grace of the guru, even a fool can become enlightened, if the person is just intelligent enough to listen to the true speech of the Guru, which is this vibration of the Guru margena. You don't have to read the scriptures, you don't have to know anything. You just have to know that the Guru is the pathway of enlightenment – that will do it.

One of the ideas of the fast path is this use of the true architecture of being. In other words, it's beyond meaning and beyond belief. It's not a system of philosophy you must master. The beauty of the Guru's pathway to enlightenment is that it is not connected to any belief system. It can appear within any belief system or outside any system. In fact the Guru Gita says that a person's mind that is full of thoughts knows nothing. It's only when the mind stops, that true knowledge arises.

Every time the throb of the Guru takes place, every time you sit with the Guru or you sit quietly and think with love of the Guru, or you recite the Guru Gita, or do sadhana in the name of the Guru, that pulsation will come out of the ocean of consciousness and that vibration will occur. You could do this sadhana on Mars – it wouldn't matter. Once you learn to listen to this vibration, this throb, the awakening will unfold.

It is oftentimes seen that, because the person's mind is captured in the phantasmagoria of the six sensory spheres, this vibration at first takes place subconsciously and unconsciously. Even though it's subconscious and unconscious doesn't mean it doesn't happen – it's just occurring unconsciously. But every time it does happen, it sends a wave of pure emptiness through the entire structure of the three bodies. All of the vibrations of the hundred fibers are stilled and pacified.

The record of our experiences is stored in the one hundred fibers and in the central, left and right channels, and along the six wheels of the chakras. But every time the vibration of the Guru margena occurs, it spreads a vibration of pure consciousness and pure emptiness.

The term "emptiness" comes from the understanding that consciousness is completely without quality. It's an impossible idea to grasp because any thought you have of it is just a set of imputed terms – and any set of imputed terms is something with qualities. At the level of thought, our apprehension of reality is always superficial and doesn't penetrate into it. This is why they say, only when the mind is stopped does the true nature of reality emerge.

When you read the Shiva Sutras, which is basically a handbook for enlightenment, or Patanjali's Yoga Sutras, which is a redux of the Shiva Sutras, you find the definition that the essence of yoga is stopping the mind. When the mind is stopped, then the true universe appears, and then the subtle throb of the Guru margena feels like the ocean of consciousness landing on your head. It really does feel like that, which says a lot about how powerful maya is. How can you miss that? Because the attention is so thinned out and diverted by maya.

Samadhi is the direct recognition of what is true and what is present. Recognition implies a perceiver. A person that is going into samadhi directly perceives the truth. When the nature of the truth is recognized, it annihilates the perceiver.

The essence of the Guru margena will pierce you, penetrate you and saturate through you. Even if this occurs subconsciously, its reality is such that it goes through all of the architecture of the four bodies.

Because it is of the nature of emptiness, whatever it comes into contact with is transformed and made empty. This is how the elimination of karma is achieved and how the karma is unspooled, layer by layer. You are dialed back to zero, although usually you don't have to get all the way back to zero. There's a point somewhere along the line where enough of the layers have been taken off so that the refinement of your intelligence can begin to come alive and you begin to truly see.

Recognition and Samadhi

The pathway of realization is a set of graduated steps. The first step of realization is faith. I'm describing something that you're not experiencing directly – but I'm describing it as someone who has experienced it, and therefore I am speaking with authority on this subject.

Once your attention is correctly directed, it begins to graduate into experience and understanding. From understanding, the effect of the Guru margena builds in your system. It wears away the covering karma and the veil of maya.

There's a point where a critical mass occurs – and suddenly you have the moment of Zen satori. You see it. Having seen it, you become convinced. As you connect to that, your own conviction strengthens. You begin to experience it directly and gain more understanding. This is where the realization within you is born. As that realization grows within you, it grows to a point of recognition.

In yoga tantra philosophy, recognition is the

union of the perceiver and the perceived, where you can no longer tell the difference between the perceiver and the object of perception. That is the condition of recognition. Realization implies a realizer and that which is realized. With recognition, they have joined together and there is no difference. I say this because I believe when attention is correctly directed, then understanding follows more swiftly.

This is why the siddha path is called the fast path because the Guru shakti takes advantage of the very structure of consciousness itself. It bypasses confusion and goes for the essential underlying truth of the nature of God within each person. It generates that vibration and forms that connection.

Each person is different. Sometimes the Guru strums that string of the innermost fiber of the Guru margena once and the person sees it right away and merges with it – and at that very moment they're realized. Another person might take a week, a year, or ten years or more. The very vibration of the innermost pathway guarantees awakening.

It is the ignition of the Guru margena that is shaktipat. Even though one has spent ten thousand incarnations building up the identity of the egocentric "I", and the assembled karma thereof, the Guru margena is so subtle that the vibration of awakening can cut through that like a laser and in an instant, grant liberation.

The movement from the unawakened condition to the awakened one is a step of extraordinary evolution. All evolution is born a struggle. So don't be thrown by the turbulence of it – it's just a part of the process. Once given, the ignition of the Guru margena can't be lost. Continue to take advantage of these opportunities for spiritual training to increase your capacities for attention, concentration, meditation, discrimination and contemplation.

One of the main focal points for your attention is to remember what's important. Keep your attention on that. The main weapon of maya is forgetfulness, the dilution of attention whether through fear or lack of effort. Remember that you have achieved the difficult-to-attain human form, wherein you are born with everything you need to awaken. It is as rare as a star at midday. Having achieved that difficult-to-attain body and even more rare, to have found the Guru and gained shaktipat after so much effort, it would just be a shame to miss the process and fulfillment of awakening because of fear or lack of effort.

The entire envelope of the physical body anchors the assembly of the soul consciousness. Before shaktipat we conceive of ourselves as a body that has a soul. But after the force of awakening, when shaktipat and the path of the Guru margena have been switched on, we see ourselves in a different light. We come to understand that we have been and always will be a soul, a component of which manifests as a physical body. As we progress in our sadhana, we become able to truly apprehend all of the elements of our being – physical, subtle physical, causal and supracausal.

This is what samadhi is. This is the fulfillment of the Guru margena. In reality, it is your first true birth because the expansion of consciousness never comes to an end. It is at this point that one begins to genuinely and directly receive reality.

CHAPTER FOUR –WHEN

The path of awakening is the purest form of mysticism imaginable as it arises within the very fabric of your being, which you experience essentially as your life.

This term 'life' is profoundly subjective. It implies the terms and basis of experience. But simply speaking, it is the life that is transformed. It is the life that is awakened. It is the life that is the field to be cultivated and changed.

When we look at life, we see a simultaneously arising set of conditions that we call stages and states. By stages I mean the stages of evolution, and by states I mean the various specific experiences that are subjectively experienced by the perceiver.

The concept of evolution up the ladder of phylogeny involves moving from simpler to evermore increasingly complex life forms. This evolution also brings with it the capacity to have a wider panorama of states of consciousness, moving from unconscious states of perception ruled by the senses to increasingly advanced states of awakening, which involve conscious awareness of the deeper and more subtle structures of your being.

From the standpoint of the spiritual panorama, the driving force of evolution is not so much the physical body but rather the complex subtle body. The human form is divided into four sheaths of consciousness, correlating to the physical body of matter, the subtle body of energy, the causal body of mind and the supracausal body of pure consciousness.

This understanding of the human form is not a random idea. Looking into reality, the masters of consciousness have seen that there are three worlds. They are the worlds within worlds. They correlate to the three divisions of the physical realm of appearance, which is the manifestation of matter (the physical body), the electrical world of pure energy, which is referred to as the kundalini (the subtle body) and the causal world of mental formation (the causal body).

There is also a fourth body (the supracausal

body) that takes the form of the Blue Pearl, a single atom, out of which arises infinite consciousness. The fourth body is one of formlessness and pure consciousness. The fourth body, pure consciousness, pervades all the three other bodies, to a totality. It's not as if, because the mind is more subtle, there is more consciousness in it, whereas matter is more fixed, so there's less consciousness in it. They are all pervaded by consciousness.

Let's look at the spectrum of evolution. The extremely complex subtle body of a human being has the capacity to transmit the electrical force of creation with incredible intensity, and can span the three worlds and the seven planes of consciousness. In other words, we are hard-wired from the firmness of the earth to the Om point – from the solidity of our physical body, to the pure energy of our subtle body, to the mental formation of our causal body, all the way to having a form that arises as infinite consciousness in the fourth body – the Om point, where consciousness merges back into its original form.

On the other side of this spectrum, let's say we

have a rock. It's mostly matter. It has a subtle body, but not a very sophisticated one. It has awareness, but not very much. It is flooded with consciousness, but there's not enough of the active ingredient of energy. Even if there's an influx of energy that goes through it, it doesn't have a complex enough subtle body to translate that influx of consciousness into awakening. It's kind of like a wave that washes over a rock. It eventually just wears it down and it becomes nothing. It dissolves into the water, then begins its movement up the chain of phylogeny. The rock itself will eventually be a conscious entity and will indeed awaken to its own nature.

So from the standpoint of awakening, this idea of stages and states is all about having a complex enough physical form that is the fruit of an ever-increasingly complex subtle form. This can translate the movement of the formless nature of mind and its perceptions into evermore increasingly meaningful states of consciousness that can more accurately reflect reality.

Spiritually speaking, most of the action in the awakening process takes place in the subtle body. This is the body that can be changed by the nature of attention itself, by the nature of consciousness itself and by the application of mind to the form.

Again, it bears repeating that this subtle body is a sub-structure, an architecture. It is the unseen support of the creation of this world system and of each individual.

There are subtle fibers, which are streams of light and are what Don Juan called "the lines of the world". You may snap into a view where the world will disappear and you'll see nothing but fibers everywhere – streaming through the trees, the streets, the cars, the people and the sky. These fibers conduct themselves through a matrix that we call the elements. The assembly of elements are earth, water, fire, air and ether or space. The panorama of the universe, both the apparent world and the spiritual universe, is supported by these countless fibers of light.

It's important to understand that these fibers of the world are not out there and we're over here apprehending them. When we truly see the fibers of the world, we see that they go from out there, to us and through us. They actually meet at all of these nexus points in the subtle body. The subtle body is the control panel, the operating system of the creation. It is the access point into the envelope of the apparent creation, where the visible material world and the subtle physical world interface with the causal envelope. As we see these fibers of consciousness, we see that they move through us in a very intense way at each of these five wheels of elemental basis: ether, air, fire, water, earth.

The subtle body is also the control center of the individual human being. It's the subtle body that connects the physical body and all its metabolic structures to the life force. The life force moves primarily through the subtle body, which is in support of the physical body. The structure of the subtle body correlates to the predominant movement and the concentration of the elements of earth, water, fire, air, and ether.

Karma, Impressions and Grace

There are three mitigating factors for the relentless momentum of the wheel of cyclic existence. This incredibly complex assembly is ruled by spiritual laws or principles that are stratified and go from base to ever more subtle and refined. As they become more refined, they become more powerful and influential. The three factors are karma, impressions and grace.

The most base force and first principle is cause and effect, known as karma, which we see as the sheer brutality of the wheel. We see this as the force of nature, food consuming food, life arising and consuming life, living, consuming life as it goes, beginning to decay, then dying and itself being consumed by life. All life forces are subject to this cause and effect.

The second mitigating factors are the impressions you carry with you as you come through the membrane of creation, from the interval known as the Bardo, through the fabric of that doorway into the apparent world. This set of conditions are attenuated to your mind and are called impressions – those feelings and thoughts that were the basis of your actions in your past life. These impressions interface with the sheer force of cause and effect, raw karma, and produce a modifying factor.

This is where you'll see a division in the conditions of a being, where one being will come into the fabric of the apparent world as a rock, or as a dog, as a cat, as a fish, as a tree or as a human. It's like a subtle equation.

When you look at it, it's impossible to describe – I'm trying to send you a picture. This is where you see this idea of evolution. What really changes are the impressions, and there is an upward rise. Not every life force comes in as a rock and goes all the way up the chain. Some do, but most of those beings have cycled off. People start in various stages of the chain of phylogeny and come in with the impressions that most suitably apply to given forms.

In this world system, the human form is one of the most rare and also most auspicious forms in which to arise. Through the inconceivably vast and sweeping energy of evolution, the human form is born with a subtle body that reaches all the way from earth, up through water, fire, air, space and into consciousness. There is a direct architecture that is present in the very form of the subtle body, that reaches from the firmness of the earth to consciousness itself.

From the standpoint of spirituality, this is extremely auspicious because you don't have to imagine something that you could never conceive of. You are born with a map in you, as you.

On the great wheel of cyclic existence, you are driven by the content of your mind to incarnate again and again, until you consciously realize the nature of the wheel. But the momentum of the wheel can be softened and mitigated by the impressions you carry forward. Just like they say, each time you cycle you learn something and you store those impressions. The fruit of those impressions gets you a higher operating body-speech-mind matrix. That higher operating matrix is proved out by your behavior.

It's important to know that this is not about understanding something in theory – it has to be proved out in action. This is why the life is important. First you are given the vehicle of the body, as well as the subtle body and the stored impressions of the mind. Then you are presented with life, which essentially is an endless series of choices that you act upon, experience the fruit of, and gain experience and knowledge from. This is an important point. It must be proved by action – it's not a white ivory tower situation. Life is war in the trenches where everything comes to the fire of proof, which of course is why life is so intense. This is the beginning stage of the awakening process.

We also have a third sphere of spiritual law. We've seen that we have cause and effect, karma, and we have the conditioning layer of the stored impressions, the lessons we've learned in one lifetime after another. You're compelled by cause and effect to be born – you don't have a choice about that, but by making good choices you will get an ever higher quality vehicle with more and more fortunate opportunity to raise yourself out of the mud and move higher and higher in what we would consider spiritual evolution.

The third layer of spiritual law would be characterized as grace. Grace is the inflowing of spiritual force, which has the ability to transmute the stew of the life force that is entangled in karma and countless impressions, both dark and light, and cause that to speed up and play out in an extremely accelerated program. This acts as a very deep spiritual push. This will happen to you in countless lifetimes. Most of the time it happens you won't be aware of it because the origin of this impulse is coming from a place that you are not yet prepared to experience consciously, thus you experience it unconsciously.

We look around at the world and there are millions and millions of species and forms of life that we can see. But the fact is, the subtle superstructure is a billion times more complicated than the physical world ever was. It has millions upon millions of forms of life that you can't see until your vision opens to the subtle realm. They are occupying the same space as we do.

What they tend to do is assemble along the elemental qualities of being, so there will be earth-based elemental beings, water-based elemental beings, fire-based elemental beings, air-based elemental beings and ether-based elemental beings. As you learn to see past the appearance of the world, you start to see the lines of the world and you see this correlation of the lines of the world with these five elemental energies.

It sounds kind of superstitious when you're saying that it's earth, water, fire, air and ether, because it's such an old-world view. But the reference is to how the lines of the world operate. They are all fibers of pure energy, but they have different magnetic energy signatures. When you see them, they form a maze that is beyond count of earth energy fibers, of water energy fibers, and so on. What's weird about it is when you see a water fiber, it doesn't look like water, it just has the signature of water – and so on with the other fibers. The apparent world of appearances is made manifest out of all of the magnetic fields of these fibers of the earth.

As you awaken spiritually, you come to see that the apparent world is just like a curtain that's been drawn over an infinitely more vast and spectacular panorama. Each of these worlds, each of these elemental spheres, is just the beginning. That's not the limit. There are countless forms of life that don't have carbon-based bodies. Their bodies are like electricity – they're based on energy, which is a form of plasma. They're conscious and may live for thousands of years. Some of them are like fish – they're just swimming around, not really knowing what's going on, just kind of drawn here and drawn there. Some are highly volitional. Some are positive and some are negative. From the standpoint of a human being, some have a positive impact on your life force, and there are many that have a very intensely negative impact on your life force.

This is also true when we get up to the sphere of mind, where there are beings that are profoundly more subtle than the elemental fibers. They live as consciousness itself, in the form of a mental formation. These beings are a thousand times more powerful in their impact on any given world system than anybody in the subtle realm because they live in a condition of pure mind.

They're at the point of origin. Much of what we call the universe is mind-born. In other words, it emerges out of the fabric of mind and there are beings that occupy that realm. In spiritual thought, what we call mind is supported by space or ether.

To complete the picture, the fourth body is pure consciousness beyond space. Beyond even space itself, there are beings that vibrate at a level of power and subtlety that do not require the support of space. They are beyond any kind of conception. These beings are in the void-space between the fourth body and the causal body. It is a field of pure consciousness, utterly formless and without quality of any kind. There's an entire sphere of beings that live there. They live in the fabric of pure consciousness. In other words, whatever occurs to them is reality. Their thought goes down through the mind, through the subtle body, into the fabric of the formation of the earth, and gives rise to a reality. Why? Because they thought it up, and they have that kind of power.

It's just like when you were a six-year old kid and walking along and said, "Oh, there's an anthill. Let's get a magnifying glass and burn all the ants." The ants haven't the slightest idea what's going on with them. You have so much power over them. It's the same idea. A simple analogy, but effective. Your entire universe may change and you have would have no idea why or what the reason was. I don't mean to generate an idea of necessary order, because it's kind of an order combined with chaos. From the standpoint of any experience you would have of it, it looks completely chaotic.

Mitigating these forces of chaos is the sphere of spiritual control known as the Siddha lineage. These are beings that were born here, went up through the chain of phylogeny, got super-complicated subtle bodies, awakened to their nature, and became masters of the mind and consciousness. They exist universally in a condition of being you can't even imagine, and they have some kind of feeling for all us ants scurrying about on this world. They would be considered the good guys, but there are also the bad guys who have the same amount of force. I'm sending you a very simplified picture to illustrate this point.

The theory of evolution according to Darwin holds that even one single evolutionary development takes 50,000 generations to change. Human beings originate from an ape-like creature: Homo erectus, Neanderthal man, all those great guys. But there are also beings that have been coming through the fabric of the Bardo from other world systems. These kinds of beings were never a rock or a tree or a fish. They went through their evolution somewhere else, and they came here for the very first time as human beings. There has been a big influx of those people, pouring in from off-planet over the last fifty years.

Grace is the third law of the creation. It is a force of high enough vibration that it can come down through the planes, impact the beings in the mental field, impact beings in the subtle body, impact beings in the physical sphere, change their karma, change their conditions, and uplift them, out of sheer intention.

I'm describing what we call shaktipat, the descent of grace. This energy comes from the Siddha lineage, who are operating on higher planes, and it is channeled to you, into you and through you. It acts as a reset. All of these spiritual forces of cause and effect, the subtle conditions that have gotten you this far, switch from a condition of production to a condition of cessation. Instead of winding on and becoming ever more burdened down, with ever more conditions and ever more compound results of cause and effect, it begins to wind off, becoming lighter and lighter.

Drawing Grace Into Our Lives

It is shaktipat that is the basis of awakening. Even though I've described this in a systematic way, it's pure mysticism. The problem is, without grace it would never work. You would never be able to get to the point of direct awakening. Well, maybe – never say never, but it would take so long.

I love watching people – you learn so much just by watching them. Each person operates from a center of a completely subjective universe, and they always do what they think is the best thing. Usually it works out good to bad, middling, or a mixture of both. What was the situation? Most of the time, they're not functioning with enough information to make a fully considered choice.

Once I got shaktipat and started meditating, I watched people even more intensely. Not just their behavior, but what their subtle bodies looked like, because that's the writing on the stone, that's the record of the soul development. The development of your soul and your spirit is written directly by the development of your subtle body – how much of it is active, how much of it is conflicted with the bound, gooey stuff of karma and conditional impressions, and how much of it is burning clear and light.

A person that has not received shaktipat tends to look like someone who fell in a mud patch. They'll be kind of clear in some places or real muddy in other places, or just a bunch of gnarly tar here and there, like an old house with windows blown out. They're leaking energy. And life is so intense that it'll find a work-around and they'll still be alive. The spirit and the raw heart – I can never get over what a person goes through.

The word shaktipat translates as "descent of grace". In other words, it's a gift. The Siddha lineage comprises a sphere of spiritual control. These are the beings that have taken station and generate this opportunity.

Shaktipat is a system of transmission. It is spoken of as a moment of truth inside a human being's arc of existence, where their positive and negative karmas come into equilibrium. This means that somewhere along the line, they have gone through enough experience, they've put their money on the table and seen it swept off the table. They've won, they've lost, and they've finally come to a point where they've won as much as they've lost.

Somewhere they started to see the pattern of it, and they've started to suspect that they were the butt of a very bad joke. Because that's what it always is – it's just out of sheer self-respect... you just don't want to be stupid anymore. It's really nothing more spiritual than that.

You'll see a person begin to investigate their choices. That is where the first recognition of the path takes place. In other words, the six poisons of ignorance, desire, anger, lust, greed and jealousy, all of the all-time favorites that are the motivators of the animal nature of behavior, have begun to be questioned. Most of those forms of behavior are animal-like. They are reflexes of instinctive existence-at-all-cost. Acting from this level, a person is willing to do anything or cause any level of suffering for what they perceive as their own self-good.

It's a moment of truth and awakening, when one

begins to give rise to ethical behavior. Simply put, the Golden Rule will never fail you: treat others as you yourself would wish to be treated. It'll never fail you. Once a person begins to replace the poisons of ignorance or desire with the higher angels of love, compassion, kindness or generosity, they recognize that there's a way out of suffering.

What they've basically come across is the first glimpse of the path. This will begin to adjust their karma away from being all bad karma, driven by the darker causality of ignorance.

While it's not necessarily illumination, it is a turning of the wheel. The path is recognized and one begins to become uplifted. This is a titanic battle. It goes across multiple incarnations, but during this time, even if one moves a half a turn towards the path, there is a sense of the well-being of the totality.

For all purposes, we can identify this turning as the recognition of the path of love. It is a path of all-inclusiveness, complete attraction and acceptance of oneself and all forms of existence. In other words, everything has to be considered.

You can't divide yourself and your perception away from all the forms of life that surround you like an ocean – to think you can is illusory and a false understanding.

The slightest turn towards this path begins the process of drawing grace. The heart will be empowered, the mind will be illuminated, and you'll begin to see the interconnections of cause and effect. You begin to test those theories out with action in your life. One positive act begins to attract grace or the power of love, which begins to uplift you many orders of magnitude more than the energy you generated to do that action.

Everybody recognizes, salutes and applauds when anybody behaves for the benefit of others at the risk of themselves. It makes the news when some passerby rushes into a house on fire, charges through the door and saves two kids. They respond: "That's a great guy". We all respect it and know it's important.

As the recognition of the results of acting out of love becomes more evident to you, you eventually draw love to the center of your heart, and you begin to act from love every day, for the sake of love itself. You give rise to love within yourself, and you notice how good it feels. You'll try it once – it's great – you'll recover for awhile – then you'll try again. Slowly but surely, the amount of times you act out of love becomes more numerous and closer together.

Grace is attracted to this force. Grace is the ultimate form of love. The more you act out of love, the more you attract that grace of love towards you. It has an incredibly profound impact on your life across the board. You change and become transformed. Eventually you come to a situation where love becomes dominant.

What is the greatest enemy of love? Simple – it's fear. Why will a person not act from love? Because they're afraid of what's going to happen. Why is a person afraid to open their heart to another? Because they're afraid that their love will be rejected. Most people risk love only if they feel love is being given to them and then only risk so much. All these different kinds of fear are what's in the way.

Bringing Karma Into Equilibrium

There's a point where the love that has had expression in your life and the fear that has produced the negative impact in your life come to a perfect equilibrium. By the time that happens, you are on the charts of the Siddha's sphere of spiritual control.

By the time your karma is in equilibrium with equal dark and equal light karma, you've been observed and somebody's got you on their project list. It's weird, but true. Because this operates through grace, what you need is shaktipat. You need to come into contact with a source of grace and receive shaktipat. This is the realm of operation of the Siddha lineage.

The entire spirit of this force is called the Guru. The syllable "Gu" means darkness and the syllable "Ru" means light. In other words, when you come into equilibrium with darkness and light, then the Guru comes into your life. That grace will come down through the planes, through the mind to the subtle body, all the way through to the body, and somehow a pathway will be cleared. Even if the shaktipat authority is 2,000 miles away from you, somehow you'll be brought together. Either the source of shaktipat will come to you, or you'll be drawn to it. You'll get shaktipat and the thunderbolt will strike. Once this occurs, everything changes.

Shaktipat is the descent of light and is called the thunderbolt. What essentially happens is that it comes down through the fabric of the highest plane, through the illumination of universal mind and the fabric of your personal mind, through the illumination of universal subtle body and the fabric of your personal subtle body, into your personal mind, speech and body. Once that happens, it's like seeing lightning in a bottle. You are the bottle, you are the vase. An explosion of lightning has gone off inside that system and everything is transformed. In that moment, the thunderbolt has stretched from the Om point to the firmness of the earth and struck you personally.

It is possible in that moment to completely and totally awaken. In fact, I am of the theory that you do completely and totally awaken, but because of the way the four bodies manifest time, it may not appear so. Once shaktipat occurs, it is impossible for you not to awaken – it's over, but the process is played out over time.

You were individually born and have gone through countless battles, which you have won and lost. Those are the pressures that have shaped you over countless incarnations. You will also awaken in a very personal way through the impact of grace, the Guru shakti and the thunderbolt. It'll be your story.

The pathway of awakening infuses light through all of the stored conditions, which are held primarily in the physical body and the subtle body. The seventeen primary nadis, the three most important of which are the ida, pingala and sushumna and the hundred fibers that come out from the chakras, are all filled with the impressions of the past.

Shaktipat reprograms all that and dials it back to zero. As it does so, the light presses in. As the space and time is cleared of past impressions, it is filled with light. Through meditation and sadhana, the fabric of your karma and conditioning evaporates and becomes filled with light.

The only variable is the degree and intensity with which you are able to bear the process of transformation. If you go at that particular speed, you will awaken at that particular time and place; if you go at this speed, you'll awaken at this time and this place. If you do this, you'll awaken in the thirtieth year of an incarnation here; if you go at this speed, you'll be born enlightened in three incarnations. It varies, but it's irresistible and must occur. Because it emerges out of the fabric of reality itself, it's not a belief system. It's not a religion. It is just the way the wheel is set up.

It's important to understand that the Guru is not a person or a place. The Guru is a piece of the creation that arose at the moment of creation and is always present. The Guru is the safety valve of the wheel of cyclic existence. Without the Guru, nobody would get anywhere – it would take too long. I'm trying to remove the superstition from a truly mystical process and give you a picture of how it works.

When you start to get same-set kinds of

incarnations enough times, then a bell goes off, and you start turning the wheel of recognition towards yourself. By off-setting fear and bringing in the higher angels of love, you gain access to the belief in the existence of a way out. Even if you don't consciously know that a way out exists, you know it unconsciously, and you'll start to assess yourself and your behavior will subtlety change.

As I said before, the point at which one's positive and negative karma come into equilibrium is the setting that draws the thunderbolt and draws the Guru. Because the Guru is infinite consciousness and infinite unconsciousness, when your ratio of good to bad karma is exactly half and half, the thunderbolt strikes.

It's a force of nature and is the way this world system is set up. It's like that because this planet is ruled by dharmic energies. It has been selected so, very consciously. It's not a matter of evolution. There are probably two or three thousand world systems that give birth to human beings. But out of those two or three thousand world systems, only the earth has the shortest distance from earth to the seventh plane to the

Om point. If you are of the human tribe, this is where you come to get off the wheel. The forces of darkness or unconsciousness, and the forces of light or consciousness are highly concentrated here.

There really is no evil, as we think of it. Evil is a religious concept. In most cases, evil refers to things that once were true, but haven't gone away. They have just hung around so long that they've become anti-systems to the present system.

Things move so slowly that you actually have different systems that are in conflict with each other in the same time and the same place, and out of some kind of cosmic joke they're antithetical to each other. That's the wheel for you, everything takes so long. Because the wheel is different at different points, people jump on here, or jump on there. But this planet is a profoundly complex nexus point, which is extremely competitive and highly concentrated. Everything happens here with great intensity and great speed.

Unspooling Karma and The Fourth State

We are aware of the fact that we're in the waking state, where the physical world is dominant. Even though the subtle energetic world is right here, our senses don't see it as much. We have to go through a de-conditioning. The mental body and the world of the mental universe are so subtle that we only know them as the fabric of the movement of our own mind, most of which we dedicate to the product of the ego, which is coarse. It's so coarse, it actually blocks out reality.

In spiritual training we are trained to actually see things in reverse. The process of meditation is one of de-conditioning the overwhelming impressions of the six sensory fields and our attachment to the physical world.

From the standpoint of reality, the only thing that is true is the fourth state, the world of consciousness. Why? Because it's always there. In the absolute sense of the word, for something to be true, it must always be true. If it's only true some of the time, then it's only relatively true, which means it's not true. In philosophy, this idea of relative truth is a way of talking about something that's very difficult to talk about. Relative truth means that it's only 20% true or 30% true, or true only under certain conditions.

We have three basic states that we're familiar with. We go through them every day, we know they exist. Why? Because we experience them. We believe them because we experience them. Very few people have experienced the fourth state. That's why it sounds kind of far out – an interesting idea, but there's very little experience of it. Because of its subtlety, you have to actively seek out the experience of the fourth state.

Thus we see that a person's incarnation will go through an arc as they gain perception of the path. They sense that there is a way off the wheel: "I don't know what it is, but I feel something". They begin to turn their behavior around and begin to act for the sake of others rather than themselves. In other words, they begin to question the primacy of the ego. They've had enough times where their ego is victorious, then was completely defeated and shattered, then it was victorious, then completely defeated and shattered. It happens enough times that they begin to genuinely doubt the existence of the ego and its worthiness as a refuge.

It turns out that the ego was a false refuge all along that never brought lasting happiness. The happiness it brings lasts only for a few fleeting seconds, then it changes. When that awareness dawns, the person has discovered the path. Once a person begins to feel that, they'll begin to change behavior and that brings the karmic equilibrium into play. The thunderbolt will strike.

The thunderbolt is pure consciousness. Pouring into you like a bolt of lightening, it strikes from the fourth body, through the mental body, through the subtle body and through to the physical body. That's it – all at once. Even one second of shaktipat is all it takes. Because the system has experienced it, now it must be accounted for. After shaktipat, what people experience is a dynamic tension because your soul has experienced something that your mind can't interpret.

So you go through a phase where you try to find adjustment. The moment shaktipat strikes, the ego is doomed. It ultimately can't survive in the same place as the pure light of consciousness. The essence of shaktipat begins the process of unravelling the samskaras and de-conditioning of those impressions.

I know everybody experiences this. I experienced it myself and you're experiencing it, where the ego is saying, "Wait a minute, those are my favorite samskaras. I didn't get the memo that I'd have to get rid of them." In that way, shaktipat is very ruthless. It's like a force of nature. It doesn't work that you get to keep your favorite samskaras, and get rid of the bad ones – they all have to go.

This is where the Guru shakti comes in because it mitigates. It tends to go for the worst first. It's kind of like an anti-toxin. It goes through and starts to root out the dark reservoirs, which are essentially various forms of fear. This is what the ego thrives on, the whole idea of separation. The ego sees the inter-connectedness of all things as the kiss of death and will hold the story in place forever: "I am separate, I am different, I am special." You go through degrees of revelation.

That quality of mitigation or tempering is actually one of the aspects of the Guru shakti. If it was just raw power, the surgery would be a success, but the patient would die. So it's a very subtle process and it produces a very selective unravelling of samskaras.

Not all of the samskaric impressions we carry are conscious. I would say the ratio is very low – the ones you're operating on consciously are just 1% and 99% are recessive. But that doesn't mean that they're not producing conditioning factors – in fact they're more subversive because they're conditioning unconsciously. The idea behind spiritual training is that slowly but surely you become conscious of the totality of the reality of the egocentric "I", which rules predominantly in the waking state. That's where the ego is the most dominant.

Why does the ego have so much force? Because it gets fed every day by the entire content of the waking state, the constant repetition of every lifetime: the emotional, psychological and experiential content of the six senses. It gains profound momentum and produces product.

We often see in our dream states that our ego becomes highly flexible and assumes a vast variety of identities. What's happening is that the recessive samskaras that usually assert themselves subconsciously are now asserting themselves consciously. The ego is highly mutable in the dream state. You can understand the necessity for the dream state because it's a safety valve for all of that unconscious pressure. This is why Western psychology has several schools of study that are based on the analysis of dreams because the subconscious is the true dynamic force of the identity. If you look for those constant themes, a pretty accurate analysis can be generated.

So we have three different world systems and three states of consciousness that hardly ever meet each other. The person we are in our dream state, our waking state, and our deep sleep state are always in intense transition and shifting. So how are we going to come to know those aspects of our own self? What tool of consciousness do we need to examine all of these relative states, from an immutable point of view that is unfailing and constant?

This is the fourth state, which is seen in the state of meditation. It is a kind of disembodied state that is not attenuated to the physical body and the six sensory fields directly, although it can explore them completely. It is not attenuated to the subtle body of sleep-with-dream and this incredible architecture of the subtle body: the winds, pranas, currents, nadis, chakras, fibers, bindhus, drops, or the subtle streams of consciousness that are the operating system of the human form. It is not attenuated to the mental body, which is without form and is outside of time and space – all-pervasive yet still relative. Even the mental body is supported by something even more subtle than itself – the fourth state.

In the bound condition we have anchored our identity to the entire force of our physical body, subtle body and mental body with the substructure of ego, intellect, memory and mind-essence. It is a false refuge and is a mistake.

And thus it's a form of ignorance guaranteed to produce increased suffering.

Because the desired outcome of any egocentric operation will never be truly fulfilling, there is a constant sensation of frustration, which is one of the strongest forms of suffering. There's no resolution and no matter what we do to try to resolve a situation, it always seems to slip out of our hand at the last second.

When we hear the word suffering, we think, "Oh yeah, suffering: getting killed in a war, getting mangled in a terrible accident, catching a terrible disease, living in poverty". They're all intense forms of suffering. But by the time those happen, it's almost a kind of resolution, it's almost a kind of weird relief. The true suffering is the uncertainty of everything – not knowing what's going to happen.

None of these positions that we are aware of – waking state, sleep-with-dream, or deep sleep state – produce a vantage point where we can get a point of view. It's not high enough ground. It's just like being half-way up the hill. Meditation is that high ground because meditation is anchored in the fourth state. The fourth state pervades the other three states. It thoroughly pervades the waking state, it thoroughly pervades the sleep-with-dream state and it thoroughly pervades the deep sleep state. At the same time, it is outside those three relative states and rests in the fabric of consciousness itself, which does not change.

Meditation is the first handle we get to begin to genuinely examine the constantly shifting sands of the three relative states. If you're trying to examine yourself in the waking state, you lose the thread instantaneously. Most days you can't keep a thread from one day to the next. The dream state is even worse.

The wise say, what we need is a vantage point of high ground – the one state of consciousness that will give you the genuine stable ground from which to find the fourth state. This doorway to access the fourth state is meditation.

Meditation is a vast subject – and that's the solution. Anybody that ever truly improved their condition, high, low or in between, used meditation as a tool. You have to. Meditation is placing your attention on the fourth state as an anchor of immovable consciousness, through which you can begin to examine all of the content of samskaric conditioning factors, and begin to gain access to the structure of yourself – physical, subtle energetic and mental.

The first rule of meditation is that you have to learn to sit down and sit still. You have to learn to sit without moving your body for as long as it takes. You sit down and you don't move for an hour. Why? Because you don't want to think about the body. You want to be freed from the body. If the mind starts racing, you see the quality of thoughts. You learn to stop reading thoughts and just see thoughts as clouds in the sky. Then you start to pay attention to the space around the clouds and slowly but surely, you can start thinning the clouds down and slowing the mind. This is all basic meditation, which everybody has to learn.

You'll oftentimes see people start the process of meditation way before they get shaktipat. These people are brave to begin to plunge into the nature of the self – without grace. Though just by putting up a fight and trying to do this, they draw grace.

Once you win grace, you draw the attention of the Guru and the attention of God and the thunderbolt strikes. Then you meditate on shaktipat itself and watch it work inside you. It will teach you about itself, just by watching it. The movement of the shakti and the movement of the kundalini are one and the same thing. The kundalini is the mother of the Guru. The Guru and God are one. The Guru didn't just invent itself. It arose as a reflex of God. It is mercy, the reflex of salvation, the way out.

What you begin to do when you start a meditation practice is you plant a flag in the fourth state. The Guru's grace helps you do this. Before you meet the Guru, you don't know where the fourth state is, but as you gain experience of it, it gives you an immovable point of view because the fourth state is pure consciousness. It does not change, it's just consciousness. Once that starts, you generate the resolve to expand the presence of the fourth state inside your life stream. How do you do this? Practice.

Shaktipat is the reset button and de-conditions each state. The mental body, the subtle body and the physical body are all reset back to zero samskaric content. Each time one of the little samskaric bits are lifted off the system, it is replaced by the vibration of the light of the fourth state. So the kundalini just slowly goes through the system very systematically purifying, awakening and illuminating.

Of all the kinds of lives you can choose, spiritual life is the most intense and the wildest you can select.

Once you gain the grace of shaktipat and the presence of the fourth state, what you need to do is meditate once every twenty-four hour cycle. You need to amplify the presence of the fourth state by sitting down and meditating for one hour, dedicating your entire being to the realization of the fourth state. Do it every day. You'll catch a glimpse of the fourth state every single time you meditate. At first it'll be just a little, but it will grow and grow because that daily hour amplifies the fourth state.

And what else? Develop mindfulness of the space between the breaths. Why? Because when you breathe in, the breath moves, the prana moves and the mind moves. When you breathe out, the breath moves, the prana moves and the mind moves. In the space between the breaths, the breath is not moving, the prana is not moving and the mind is not moving. By placing your attention on the space between the breaths, you lean into silence and that will cause it to become amplified.

As you lean into the space where the mind is slowing and stopping, you will actually begin to experience profound silence at an ever-increasingly rapid rate of speed. We call this mindfulness. What are you being mindful of? The fourth state. This repetitive attention to the fourth state in formal meditation produces an impression of the fourth state. By repetitive practice you have waking state, sleep-with-dream, deep sleep, and now you've got a fourth state, which is of the nature of pure consciousness.

As you go through this process there's a point where your meditation begins to gain force and color your every experience. When that starts to happen, it starts to infuse itself inside the six sensory fields. Silence starts to appear sensorially. You start to see behind the event of things. You see behind cause and you see effect, as well as the substructural support of both. And you see something even more profound – you see not-doing. You see nothingness or emptiness supporting both cause and effect.

Because you're dealing with a very profound intangible, you're teaching your system a new state. Think about when you're born. You're crawling around in this new world of sounds, spaces and sensations. There's this incredible thing – the green chair in the kitchen. Then one day you finally just see it as a chair. The same thing happens with the fourth state. It kind of sneaks up on you. You begin to see and experience it in a tangible way.

There's actually a term for the operation of the slow arising consciousness of the fourth state, and that is "witness". There is a part of you that was always conscious, was always enlightened and was never unenlightened. That is also known as bodhicitta, the awakened nature, but it's been so veiled over by the operations of the production of the wheel of cyclic existence that it's been lost into the deepest unconscious memory.

The gaining of shaktipat, the descent of grace, which clears away oceans of karma by the second, is an incredibly transcendental event. Shaktipat rocks – world systems of karma vanish by the second. You need that rate or you'll never really get there. Your daily practice, meditating every day for that hour, produces a conscious awareness of the fourth state. You're prying open a place for the fourth state to exist in the panorama of your own universe. It was always there, but you forgot about it.

After you go through a certain amount of spiritual training, everything actually goes in reverse. It is the fourth state that is real. The appearance of the world is like a shadow play across it – like a movie, a play of light and color.

The daily practice is planting that banner of enlightenment. You're signifying to all the other parts of yourself that this is being declared important and you're paying attention to it. In your daily practice of meditation, you learn to generate awareness of the space between the breaths, where the fourth state reveals itself. Every time you breathe in and out, the fourth state, complete silence, reveals itself for that one second between the two breaths – every time. Of course it's happening unconsciously, so it doesn't have much power, but as soon as you pay attention to it, you start to make it conscious and it becomes amplified.

It's just like how the child learns to see the chair as the chair, or how it learns to walk and begin to identify the world. The awareness of the fourth state begins to develop in the same way. By going there each day, it flows in with greater and greater force. It's always more because it's really an ocean pouring into a bottle.

You really are – and always have been – the fourth state. Everything has actually emerged from there, not the other way around. It seems like that now, but once you get there, everything reverses. All the relative world just goes away and you are eternally present in the ever-present now of the ocean of consciousness.

From the first moment of shaktipat to your very first meditation, to the swimming upstream of the constant illumination of witness, to the day you break through to the direct apprehension of the fourth state, is actually a very short amount of time. Your first direct apprehension of the fourth state is called samadhi – the intuitive apprehension of the highest truth – and this is the fruit of shaktipat.

The greater the yearning and the intensity of mindfulness that one has for enlightenment, the greater is the intensity of the unspooling process that will take place. Your life will begin to change in increasingly-accelerated ways. This is the positive benefit of shaktipat. It is useful to understand that this is happening so that one can maintain an equilibrium through the increasingly dynamic changes that follow.

This is a practice that can be generated in virtually any state. It can be generated in the waking state, as you move throughout your day and your busy lives. It can be generated in the dream state. You should also try to directly enter the fourth state at least once a day. Sit down and meditate. Give rise to the SoHam. Follow the breath in. Press your attention into the space between the breaths. It is very potent in relationship to the breath component of your Hatha practice, and it will kindle and increase the kundalini in your system.

I wish you all a life of joy and awakening. And always remember, God dwells within you as you.

GLOSSARY

A – Ka – Tha ~ Within the cerebrum of the brain are many aspects of the subtle body. The triangle referred to in the Guru Gita is a formation that is in the physical brain as well as the subtle body. Inside this triangle is absolute void, infinite black consciousness, without form and without quality. This triangle is located in the center of the brahmarandhra and is formed by the Sanskrit alphabet. Each of the letters of the Sanskrit alphabet are understood to be sacred mantras. The first letter of the alphabet is "A" (pronounced ah). "A" and the next 15 letters form the first side of the triangle. The next set of letters beginning with "Ka" form the second side of the triangle and the last set of letters beginning with "Tha" (pronounced ta) form the third side. A, Ka and Tha are at each of the three points. The Guru's two lotus feet, which are Ham and Sa are in the center of this sacred triangle.

Ajapa Japa ~ Japa means repetition of a mantra. When 'a' precedes a word in Sanskrit, it gives negation to the following word, so ajapa means non-repetition. Ajapa-japa means the nonrepeated repetition of a mantra – in other words, a spontaneous repetition, such as the natural sound of the breath. A mantra is a sacred syllable used as an object of attention in meditation. By lightly holding a focal-point, such as a mantra or attention to the breath, the mind is able to stay alert as it naturally sinks into deeper levels of silence.

Atman ~ The Self; the presence of the Divine within the individual. The Soul.

Bardo ~ In Tibetan, this literally means "the interval". There are usually six bardos that are spoken of. In the context of this book, Mark is referring to the bardo thodol, the intermediate state that enlightens upon contact. The six bardos are: the bardo of the waking state (the interval between birth and death); the bardo of the dream state (the interval between going to sleep at night and awakening in the morning); the bardo of meditation (the interval between the breaths); the bardo of death, chikhai bardo (the interval between life and the afterlife); the bardo of luminosity, the chonyid bardo (the interval between death and rebirth); and the bardo of rebirth, the sidpabardo (the interval between the afterlife and the new experience of the waking state). The bardo thodol comprises the three last bardos, chikhai, chonyid and sidpa, as it is possible to spontaneously gain enlightenment during any of these states by direct recognition of reality.

Bindhu ~ The seat at the back of the skull. Of all of the seats and places in the subtle body, the third eye is the easiest point to control everything in the subtle body because it is both conscious and unconscious, dark and light. It brings into equilibrium the left and right side, ida and pingala, as well as the void essence between formless consciousness and consciousness with form. With most of the chakras, energy goes in and out on the same current but at the throat it's like a highway that is divided – like hitting a "Y" in the road, a soft left or soft right. One part goes straight up the spine, up the back of the head, engaging this seat called bindhu at the back of the head, over the crown of the head and down into the ajña chakra in the third eye. When it comes down to the third eye, there's also a current that goes straight from the back of the head to the third eye. The two lobes of the brain are on that axis and it's just a core of nerves through that center piece. The kundalini will start firing from the back of the brain, straight to the forehead, and open these nadis. When that occurs, it's like a seam opens at the back of the head and it kind of unfurls and opens at the front of the head. It's extremely blissful.

Blue Pearl ~ A scintillating blue light or bindu, the size of a tiny seed that can appear during meditation. It is said to be the doorway to the inner Self and contains the entire universe. It is the form of the fourth body. It is also known as the rupa in the Guru Gita.

Brahmarandra ~ Literally "the hole of Brahman". Located in the thousand-petaled chakra at the crown of the head, it is the soft spot at the brain, found in infants. It is the point where the infusion of life takes place in the human being. It is just underneath the skull embedded in the upper layer of the cerebrum, which is the part of the brain that is dedicated to the highest brain functions. This is also the seat of the Guru. It is the superior place to exit the body upon death.

Causal ~ The third body of mental formation.

Chakra ~ Literally: wheel. They are axes of energy within the subtle body and there are seven of them. Six are relative and one is absolute. The relative chakras are at the forehead, the throat, the heart, the navel, the genitals, and at the base of the spine. Each of the chakras are resting on the sushumna, like decorations on a Christmas tree, and they have channels that flow out from them. At the base of the spine there are four channels, also known as petals. At the genitals there are six. At the navel there are ten. At the heart there are twelve. At the throat, sixteen. At the forehead there are two.

Dharma ~ That which upholds, supports or maintains Law or Natural Law. As well as referring to Law in the universal or abstract sense, dharma designates those behaviors considered necessary for the maintenance of the natural order of things, and may encompass ideas such as duty, vocation, religion and everything that is considered correct, proper or decent behavior.

Guru ~ From the Guru Gita: "The word Guru is composed of two sacred syllables. Gu, which represents the darkness and Ru, which represents light. Gu is maya and Ru is the destruction of maya. Gu is the state which is beyond the three gunas and Ru is emptiness. The Guru is he who gives the experience of darkness melting into light, maya dissolving into clarity and formation revealing wisdom." The Guru is the force of the universe which provides a bridge from the darkness to the light, which may be embodied in a human being or other sentient form.

Guru Gita ~ The Guru Gita is the core section of 352 sutras in the latter portion of the ancient Indian text known as the Skanda Purana. Gita means song, and indeed these sutras are a song in praise of the Guru and in recognition of the power of contemplating the Guru's nature, especially through the vehicle of the repetition of these verses. Mark Griffin has selected 108 of these verses for the use of serious seekers - those who are sincerely interested in spiritual training. The daily recitation of the Shri Guru Gita is one of the practices of the Hard Light Center of Awakening.

Guru Principle ~ Also Guru Tattva; The Guru is understood to be more than a particular person, but is rather a principle of creation and a ray of the Creator, designed to provide the agency of grace, which acts as a doorway for awakening. There is said to be an outer guru and an inner guru.

Hamsa ~ Hamsa means swan in Sanskrit. The swan is a symbol for the soul. Many great saints and siddhas, such as Baba Muktananda Paramahamsa share the name 'Paramahamsa', meaning great swan or great soul. Also, see SoHam.

Jñana ~ Knowledge. The Guru principle flows into the mind and begins to burn off karma and confusion, and bring understanding and recognition.

Iccha ~ Will. The will of God is transmitted directly into you in shaktipat and you are brought into alignment with the Divine Will.

Ida ~ Known as the moon nadi, this channel or prana current ascends up the left side of the central channel, the sushumna. It is red in color, and is female or yin. It is connected to the parasympathetic nervous system and has a calming and cooling effect on the mind when it is activated. It originates at the base of the spine and terminates at the left nostril. The ida is essentially empty in nature and is of the reflective quality of mind.

Kali Yuga ~ Yuga means an age or a cycle of time. There are four yugas in a day of Brahma, which lasts 4,320,000,000 years. Our current age is Kali, which is a period of chaos and entropy, relative to the other yugas. The four yugas are: Satya Yuga, the Golden Age; Treta Yuga, the Silver Age; Dvapara Yuga, the Bronze Age; Kali Yuga, the Iron Age (present time).

Karma ~ "Action" or "deed"; that which causes the entire cycle of cause and effect.

Kundalini ~ Literally "coiled one". The primordial Shakti lies dormant, coiled like a serpent, three and a half times, at the base of the spine in the muladhara chakra. Once this mystical energy or life force is awakened, it then ignites the quest for spiritual knowledge in the seeker. Through shaktipat, a Siddha master is able to activate this divine cosmic energy so that it can arise and expand consciousness through the purification that it brings about.

Maha ~ Sanskrit meaning great, large or massive. Maha chakra means great chakra, the sahasrara.

Mantra ~ A sacred word or energetic vibration, which leads the mind inwards towards increased stillness and emptiness.

Margena ~ From the Sanskrit root maarg "to seek, to strive". Translates as the way or the path and is used here in conjunction with Guru, as in Guru margena, to denote the path of the Guru. The path of the Guru really has three meanings: One – the instructions given by the Guru for the specific practices of spiritual training that he recommends. Two – the connotation that Guru yoga itself is the swiftest road to realization. In verse 2 of the Guru Gita, Parvati asks Shiva, "Kena Margena" – which path shall I take to awaken? And three – the inner road that the Guru awakens, between the inner heart and the crown of the head; the key section of the sushumna.

Maya ~ In the passing between the two layers of consciousness there is a kind of frequency of forgetfulness, a loss of knowledge of Self, and thus a cycle is generated from the beginning of the creation of this and that, the ocean of consciousness and the creation of the manifest universe. This forgetfulness arises as a subtle form of ignorance. In Eastern philosophy it is referred to as maya.

Nadi ~ Subtle nerve or fiber. The subtle or astral body consists of a very complex psychic network of 72,000 subtle fibers, interconnecting the chakras in the subtle body. The life force, the prana, circulates through the nadis. The sushumna, ida and pingala are the three principle nadis.

Om Point ~ Absolute ground zero for reality at a universal level. There is only one Om point in each person and it pervades the totality of everything. The impact of the Om point is to transform everything it touches into the fabric of reality itself. It is beyond quality or condition and is pure consciousness.

Paramatma ~ Para translates as supreme, most

high, ultimate, perfect. Atman is the Self, thus paramatma refers to the supreme universal Self.

Patañjali ~ see Yoga Sutras.

Pingala ~ Known as the sun nadi, this channel or prana current ascends up the right side of the central channel, the sushumna. It is white in color and is male or yang. It is connected to the sympathetic nervous system and has a heated or energizing effect on the mind when it is activated. It originates near the base of the spine, between the first and second chakras, and terminates at the right nostril. The entire alphabet of the creation is suspended inside the pingala. It is the expression of the creation of the world.

Prana ~ The vital life force energy that sustains the body. It animates all physical and material forms including the human body and is absorbed into the body through the breath. There are five principal forms of prana: the rising force known as prana; the descending force known as apana; the cyclical revolution force known as samana; the splitting and attracting force known as vayana; and the force of infusion known as udana.

Rupa ~ Literally: form. See Blue Pearl.

Rupatitam ~ Beyond rupa, beyond all formation. Emptiness.

Sahasrar ~ Literally "thousand". The thousand-petaled spiritual chakra at the crown of the head. This seventh chakra is the gateway to the highest states of consciousness and is considered to be the abode of Shiva.

Samsara ~ Literally "continual movement". The cyclic existence of reincarnation: birth, death and rebirth or transmigration. This word points to the idea of being stuck in the relative ever-changing aspect of creation without experience of the absolute non-changing reality. The conditioned endless karmic cycle of worldly existence that is transcended once one achieves the highest state of enlightenment. See The Wheel of Cyclic Existence.

Samskara ~ Impression, gained from the residual effect of actions, feelings and encounters that we are storing from this life and countless past lives. Samskaras are visible to one with an awakened eye as though they were little grains of rice on the strands of the nadis, the subtle nerve fibers of the pranic body. Negative impressions show as dark grains and positive impressions as whiter grains. In either case, it is the residual samskaras that act as impurities in the system and create activity in the subtle body, and thus keep the mind active. To fully perceive the Self, the mind must be completely still so it is critical that these impressions are purified from the system. This is achieved through the grace of the Guru, through meditation and through conscious breathing.

Sangha ~ Sanskrit meaning "association", "assembly," "company" or "community" and most commonly refers to the community of spiritual seekers.

Shakti ~ Power, force. Guru shakti translates as the power of the Guru. Iccha shakti translates as the force of will. Prana shakti is the force of the prana. Shakti also refers to the Goddess Shakti, the consort of Shiva. It is also the life giving force, the potency of the female energy, the creative principle and its expression. Through training with a true Guru, one's spiritual energy or shakti builds up or accumulates, gradually empowering the seeker with the ability to realize the truth.

Shaktipat ~ Shakti means "energy" and pat means "descent or falling down". The transmission or descent of grace from a Guru to his disciple through touch, sight, sacred word or thought. Shaktipat activates the dormant kundalini in a person who is open to receiving it. This transference of energy from the Guru to the disciple is known as the bestowal of grace. Shaktipat is a gift given by the Guru.

Shiva ~ Literally "good or auspicious". The ruling force in our current world system, the dominant law and the primordial Guru. The God of creation, regeneration and rejuvenation as well as destruction and annihilation, Shiva is often represented sitting on a tiger skin, holding a trident with snakes coiled around his neck and arms. He is shown sitting in meditation with his third eye open, hence the reference in the Guru Gita to Shiva with his three eyes.

There are two aspects to Shiva: the Shiva who is transcendent and pervades everything, known as para-Shiva; and the Shiva who is immanent and exists in this world – the apara-Shiva. It is the para-Shiva, the transcendent Shiva, who is the author of the five acts. In order to perform the first three of these acts, he expresses himself as Brahma, Vishnu and the apara-Shiva. Brahma is responsible for the creative impulse; Vishnu is responsible for sustaining creation; and Shiva is responsible for destruction, bringing each phase of creation to a close.

Kashmir Shaivism holds Shiva to be the pre-eminent God of all Gods. In his capacity as supreme Godhead, Shiva is recognized as the author of the Guru Gita. Mark Griffin is in the Siddha Lineage, which is a Shiva lineage that has its roots in the mystical tradition of Kashmir Shaivism. The Guru Gita springs from this tradition as well.

Shiva Sutras ~ The Shiva Sutras are a collection of seventy seven aphorisms that form the foundation of the tradition of spiritual mysticism known as Kashmir Shaivism. They are attributed to the sage Vasugupta of the 8th century C.E.

Siddha ~ A master of consciousness. Literally means accomplished; one who has accomplished the goal of life – achieving full realization, liberation.

SoHam ~ Said to be the natural sound of the breath moving in and moving out. At different stages of sadhana you may experience the So and Ham reversing. SoHam is the same as the Hamsa mantra; the syllables are switched reflecting the nature of the mantra to reverse. Literally the syllables translate to I Am That (Sah = that, and Aham = I), where That is understood to be the Infinite Ocean of Consciousness. Ham is the pool of energy at the crown of the head. So is the pool of energy at the base of the spine.

Supracausal ~ Beyond the causal; the fourth state of consciousness; the fourth body.

Sushumna ~ The central prana channel through which the Kundalini rises. Starting at the base of the spine where the Kundalini serpent power is coiled, the sushumna rises up through the center of the body to the crown of the head. This subtle principle nerve is the only nadi that connects the first six chakras with the seventh chakra at the crown of the head. It, along with the ida and pingala nadis, are the three principle nadis of the subtle human body.

Wheel of Cyclic Existence ~ Also known as samsara or the Twelve Spokes of Dependent Origin. It describes the journey from birth through death to rebirth. The cycle goes from ignorance > dispositions > consciousness > name and form > six sense fields > contact > feeling > desire > appropriation > becoming > rebirth > aging and dying. Ignorance results from believing you are the body and knowing only the three states of relative consciousness: waking, dreaming and deep sleep, with complete ignorance of the fourth state of consciousness, the transcendental ocean of consciousness. This ignorance gives rise to the condition of dispositions, which is interdependent cause and compound effect. This in turn gives rise to consciousness, which is a sense of singular identity and the beginning of the ego. As identity is formed, we then see the development of name and form, which represents a separation of "I" and "That" and we begin to define ourselves as separate perceivers. This perception occurs through the development of the six sense fields (sight, sound, taste, touch, smell and intuition). With the development of the six sense fields, we then begin to reach out and touch the universe. This is the quality of contact. As soon as contact is generated, it gives rise to a condition of feeling and this produces a psychic imprint of sensation. These in turn lead to desire or grasping. Having experienced something, we try to hold on to it. This gives rise to the condition of appropriation, which is the process of saying, "This is my experience". It is attachment. The cumulative effect of all this emotion, desire and grasping, leads to a large collection of "baggage", which we have become and are now identified with. The fruit of this mass of identity consciousness seeks a form dedicated to repeating those experiences again and again, and thus we have rebirth. As soon as we are born, we immediately begin aging, decaying and dying. We experience this death as suffering, but the wheel of cyclic existence shows us that it is really just a result of one condition having led to the next condition, which leads to the next condition, and so on. The intervention of the Guru allows this cycle to be broken at any number of spokes of this wheel, which frees us from the cyclic nature of samsara.

Yoga Sutras ~ 196 sutras summarizing the essential tenets of yoga, authored by the great sage Patanjali in the 2nd century BCE. In the Yoga Sutras, Patañjali prescribes adherence to eight "limbs" or steps to quiet one's mind and achieve liberation. The Yoga Sutras are divided into 4 chapters or books (known in Sanskrit as pada), divided as follows: Samadhi Pada (51 sutras) – Samadhi refers to a blissful state where the yogi is absorbed into the Self. Sadhana Pada (55 sutras)– Sadhana is the Sanskrit word for "practice" or "discipline". Vibhuti Pada (56 sutras) – Vibhuti is the Sanskrit word for "power" or "manifestation", or siddhis (supra-normal powers). Kaivalya Pada (34 sutras) – Kaivalya literally means "isolation", but is used in the Sutras for emancipation, liberation or moksha.

Yoga Tantra ~ The combined body of wisdom encompassing the teachings of how to get enlightened, including religious and mystical literature and the cognitions of the seers gained from direct perception. Yoga literally translates as union. It is a science, not a religion. Tantra means loom, and refers to the weave of the fabric of reality.

ADDITIONAL RESOURCES

The Hard Light Center of Awakening offers a forty volume curriculum of study of the teachings of Mark Griffin, entitled **Deepen Your Practice**.

Each month you receive both audio and written material directly from Mark's talks. You also receive an introduction to the topic of the month with an explanation of how it fits into the overall curriculum and scope of what you are studying, a glossary of the Sanskrit or technical terms that Mark has used in that month's teaching, and a few questions for you to contemplate during the month.

These lessons are presented in a sequence that build up an entire course of study in Spiritual Training. Month by month, he will take you further along the path to your own Awakening.

The course is available as a monthly subscription program, which can be mailed to you as a wire bound booklet + CDs, or digitally downloaded as a PDF and/or kindle ebook + mp3 podcasts.

For more information, please visit: www.hardlight.org/deepen_practice.html.

Individual Deepen Your Practice modules are also available (written format only) via the Amazon.com web site as kindle ebooks. Search "deepen your practice griffin" for a complete listing of all 40 volumes.

The Hard Light Center of Awakening website also offers free podcasts and video selections of talks that Mark Griffin has given, as well as inspirational quotes, and a calendar of events.

The Guru Gita mentioned throughout this text is available as a kindle ebook, a paperback book or hardcover book from Amazon.com or the Hard Light Online Store.

The Guru Gita is also available as an app from the iTunes App Store. Search "guru gita awakening", or visit: http://itunes.apple.com/us/app/guru-gita-the-essential-text/id448482878?ls=1&mt=8

Other Books by Mark Griffin

The Essential Spiritual Training Series

Volume 1: Shri Guru Gita

Volume 3: Spiritual Power

Volume 4: The Bardo Thodol
A Golden Opportunity

Volume 5: 108 Discourses on Awakening

These titles and more are available through the Hard Light Online Store: visit www.hardlight.org/store/ or through Amazon.com

The Hard Light Online Store also carries over 100 CDs featuring guided meditations and talks, as well as transcripts of many talks.

www.ingramcontent.com/pod-product-compliance
Lightning Source LLC
Chambersburg PA
CBHW051428290426
44109CB00016B/1475